The Civil War Diary of George R. Farley, 117th New York Infantry

Edited by

James S. Pula

ISBN 978-0-941504-03-4

WinStan Press
Academic Consulting
Washington, DC

The cover image is Sergeant George R. Farley during the war.

Dedicated to all of the

Men from Oneida County

who Volunteered to Save the Union.

147th N.Y. VOLS.

Contents

Introduction

The Civil War was the most destructive war in history in terms of American deaths, property destruction, and the lingering sectional and racial divisions that remained after the official end of hostilities. At its most basic element, it was the collective actions of literally millions of people who took part in these events that shaped the future destiny of the nation. Fortunately for the historian, a large number of the participants left memoirs, letters, and diaries describing their hopes, their fears, and their experiences. These original sources form the building blocks on which historians attempt to explain what actually happened and the results. One of the thousands of these sources was a diary kept by George R. Farley, a member of the 117th New York Infantry.

A student and clerk from Prospect, Oneida County, New York, Farley's journal is at times frustratingly lacking in the kind of personal observations that historians value for lending the human element to their writing; after all, history is the story of people and the incidents in their lives is what brings their stories to life for later generations. Yet, despite the often abbreviated entries, his diary is important because he served in some of the less studied Civil War campaigns about which fewer sources survive – the Suffolk, Virginia, campaign of 1863; the siege of Charleston, South Carolina, in 1863-64; the Bermuda Hundred campaign of 1864; and Gen. Benjamin Butler's advances on Richmond and Petersburg in 1864. As such, it provides some detail to the timeline for those events, as well as occasional descriptions and a view to what was of interest to the common soldier.

The editor wishes to thank Donald Wisnoski, the Oneida County History Center, the United States Army Heritage Education Center, and the Library of Congress for the illustrations used in this work. He also wishes to thank Hal Jespersen for the fine maps presented to assist the reader.

Camp Huntington in Rome where the 117th New York trained.
Library of Congress.

40th ANNUAL
REUNION
117th N. Y. Vols.
SUMMIT PARK
1902

Off to Save the Union

August 18, 1862 — April 15, 1863

Like many residents of Oneida County in the time of westward movement, the Farley and Jones families were relatively recent arrivals to the area. Rev. John Farley moved from Connecticut to Prospect, New York, sometime around 1825. The family of Ann Jones immigrated from Wales in 1800, residing in Philadelphia before moving to a farm near Remsen in 1806. Darius and Ann married on May 31, 1836, with George R. Farley coming into the world in Prospect on December 3, 1840. When Southerners in Charleston, South Carolina, fired on Fort Sumter in April of 1861, igniting a national fratricide, George was a twenty-year-old student and clerk living in Prospect.[1]

Men rushed to the colors to preserve the Union, Oneida County contributing the 14th, 26th, and 97th Volunteer Infantry Regiments, known locally as the First, Second, and Third Oneida. Despite the enthusiasm, the first fifteen months of the war did not go well for the Union's forces in the east. Defeat followed defeat at Bull Run, in the Shenandoah Valley, and on the York Peninsula in Virginia with no substantive progress subduing the rebellion to show for the ever-lengthening casualty lists. Amid the growing despair, on July 1, 1862, President Abraham Lincoln called for 300,000 additional volunteers, with New York's Governor Edwin D. Morgan issuing his own challenge the following day: "This appeal is to the State of New York: it is to each citizen. Let it come to every fire-side. Let the glorious example of the Revolutionary period be our emulation. Let each feel that the commonwealth now counts upon his individual strength and influence, to meet the demands of the government. The period has come when all must aid. New York has not thus far stood back. Ready, and more than willing, she has met every summons to duty. Let not her history be falsified, nor her position lowered."[2] *Among those who answered the appeal for the Fourth Oneida was George R. Farley.*

Destined to be mustered in as the 117th New York Volunteer Infantry, the men assembled for their training at Camp Huntington in Rome where they were introduced to their new commanding officer, Col. William R. Pease. A native Utican who graduated from the U.S. Military Academy at West Point in the Class of 1855, he was serving as a first lieutenant in the 7th United States Infantry when the war broke out. Assigned by the War Department as mustering officer in Cincinnati in May 1861, he quickly gained promotion to captain

---

1 Daniel E. Wager, *Our Country and Its People: A Descriptive Work on Oneida County, New York* (Boston: Boston History Company, 1896), Vol. III, 11; Ann Jones Farley, https://www.findagrave.com/memorial/103279192/ann-farley; Darius Farley, https://www.findagrave.com/memorial/20499674/darius-farley.

2 James A. Mowris, *A History of the One Hundred and Seventeenth Regiment, N.Y. Volunteers* (Hamilton, NY: Edmonston Publishing Company, Inc., 1996), 14-15; Samuel W. Durant, *History of Oneida County, New York* (Philadelphia: Everts & Fariss, 1878), 643.

the following month with an assignment as mustering and disbursing officer for Central New York headquartered in Utica. Upon his arrival, Gov. Morgan tendered him the rank of colonel charged with organizing, drilling, and otherwise preparing the new command to take the field.[1]

Farley enlisted in Trenton as a private on July 12, 1862, mustering in as a corporal in Company F on August 13.[2] *He began his diary the day he left Prospect for the regimental rendezvous in Rome. Once mustered into service, the regiment was assigned to the defenses of Washington until April 1863.*

Col. William R. Pease.
Donald Wisnoski.

Prospect Aug 18th 1862. This morning started for Rome, went with George Lamb,[3] arrived there at 11 o'clock A.M. went to the Barracks and stayed until night, then went to Willett's hotel. Stayed all night, has been pleasant.

Tues. Aug 19th 1862. We found ourselves all right, went to the Barracks. Eat breakfast received Twenty five Dollars of Government bounty.[4] Two Dollars for enlisting. Stayed on the ground that night, it is pleasant.

Wed. Aug 20th. We were mustered as a Regt. Warm.

Thurs. Aug 21st. Today I received thirteen Dollars for month's pay. Fifty Dollars Town bounty and Fifty Dollars State bounty making in all One hundred and Forty Dollars. Father and Garret was here today. I

---

1 Mowris, *History*, 14-15.

2 George Fairhead, Annotated Muster Roll of the 117th New York, Oneida County History Center, Utica, NY.

3 Henry Lamb was a 21-year-old farmer from Trenton who served throughout the war. Farley later corrects his first name to Henry.

4 By summer 1862, bounties were being offered by federal, state, and local authorities to encourage men to volunteer. Often when added they could amount to a year's salary for the average unskilled worker.

sent home One hundred and Twenty three Dollars. Took Twenty five with me. Very pleasant.

Aug 22nd 1862. This morning we received orders to be ready to march at 11 o'clock. We marched to the [railroad] cars, started from Rome at 1 o'clock P.M. Stopped at Utica 15 minutes, arrived at Albany 10 o'clock P.M., took a steam-boat.

Sat Aug 23rd 1862. In the morning we found ourselves half way between Albany and New York. To-day we received our clothes, sent our old ones in a box directed to L. D. Coon Trenton. At 5 o'clock we arrived at the Jersey shore, took the cars, rode all night.

Sunday 24th 1862. Found ourselves in Philadelphia at four o'clock we had a breakfast. Henry Lamb was taken sick and we left him here. We arrived at Baltimore at 5 o'clock. We marched through the city, ate supper, took the cars for Washington.

Aug 25th 1862. At four o'clock A.M. we arrived at Washington. Here I had my likeness take, at ten o'clock we marched to long bridge, and we could not cross for the bridge was broke. The distance is 2 miles. We came back went to Georgetown which is two miles and up to Tenallytown. We arrived at 3 o'clock P.M. We made our beds on the ground tonight. The weather is fair.

Tues. Aug 26th. We have been in camp all day. Pleasant.

Wed. Aug 27th. Drilled two hours today, lay around the rest of the time. It is pleasant.

Thurs. Aug 28th. To-day we rec^{d} our tents. Pleasant.

Friday Aug 29th. At one o'clock A.M. we were called into line with our knapsacks and marched to chain bridge. We got there at day break, five miles. Stayed there until noon and then we started for camp. When we got there we had orders to go on picket. I was on picket about five miles from Tenallytown.

Sat Aug 30th. We are on picket this morning at four o'clock we were relieved. Started for camp. While we were gone the men moved our things to Fort Pennsylvania, only a few rods. Is pleasant, we camped here all night.

Sunday Aug 31st. This morning it rains. At noon it cleared off pleasant.

Monday Sept. 1st. This morning we had orders to march to Fort Alexander, which is five miles. We started at noon, arrived there at four pitched our tents. Just got to bed when we had orders to go and sleep in the Fort. It was cold here but pleasant.

Tues. Sept 2nd. Today I was corporal of the guard. Pleasant.

Wed Sep. 3rd. At nine o'clock I came off guard. I have been in camp the rest of the time.

Thurs. Sep 4th. Drilled 2 hours, the rest of the time lay in camp. It is pleasant.

Friday Sep 5th. Today our Rgt. have been chopping down trees. We do not hurt ourselves. It is pleasant.

Sat. Sep. 6th. Stayed in camp today, the weather keeps fair.

Sunday Sep 7th. We had a General inspection. It is pleasant.

Monday Sep 8th. To-day our Regt have been chopping. Pleasant. We stayed here until Wed 10th 1862, then we moved to Camp Mud. We worked on Soapstone Fort.

Nov. 24th 1862.[1] We moved to Camp Morris, here we worked on Soapstone Fort and Fort Pennsylvania.

Wed. Dec 24. 1862. We moved to camp near Fort Ripley, connecting Fort Ripley, Fort Franklin and Fort Alexander.

Jan 17th 1863. We were paid. I rec^{d} $30.20.

Feb. 26th 1863. I was appointed Sergt. and rec^{d} my Sergt stripe.

Sat. Feb 28th. We were mustered for pay we have four months pay due

---

1 There are two unexplained gaps in the original diary between September and November and then again for a month from November 24 to December 24.

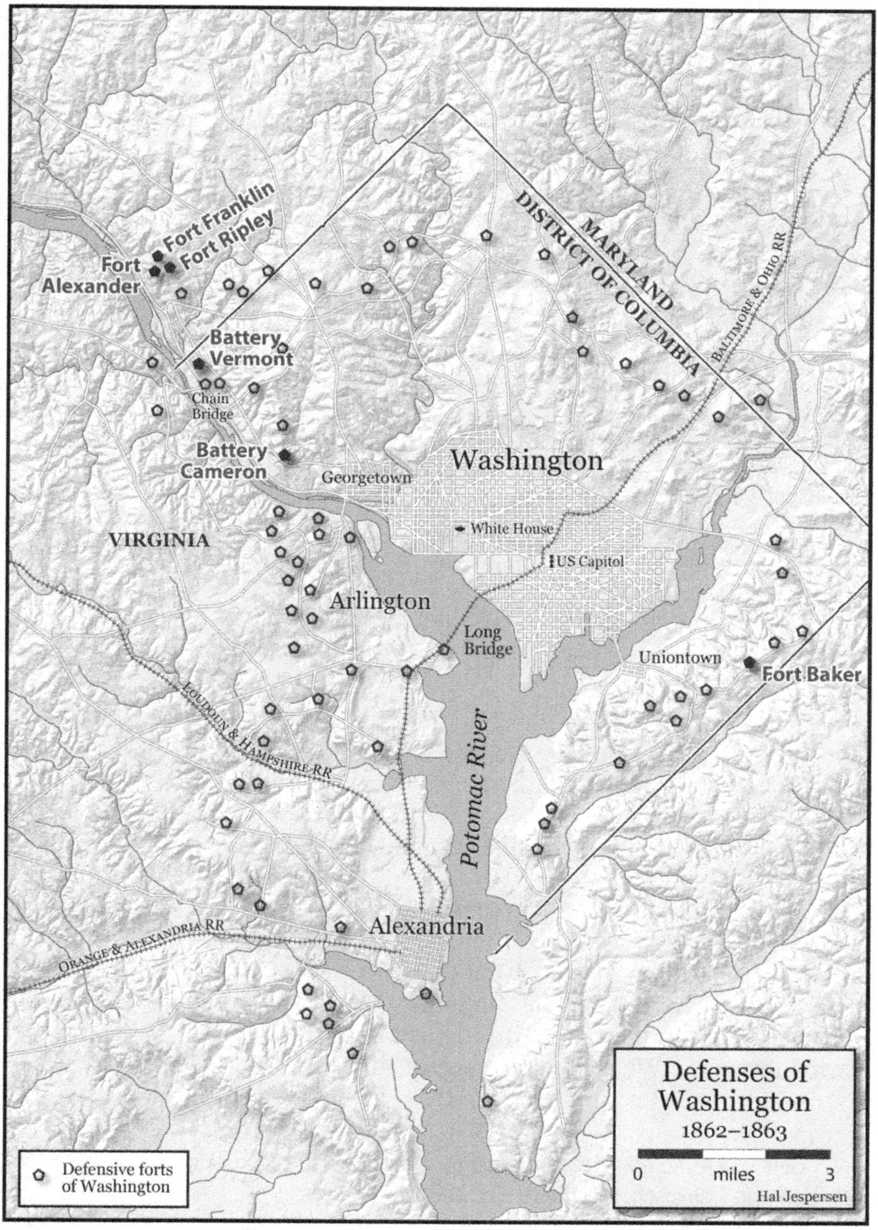
Fort Franklin
Fort Ripley
Fort Alexander
Battery Vermont
Chain Bridge
Battery Cameron
MARYLAND
DISTRICT OF COLUMBIA
BALTIMORE & OHIO RR
Washington
Georgetown
White House
US Capitol
VIRGINIA
Arlington
Long Bridge
Uniontown
Fort Baker
LOUDOUN & HAMPSHIRE RR
Potomac River
Alexandria
ORANGE & ALEXANDRIA RR
Defenses of Washington
1862–1863
0 miles 3
Hal Jespersen
Defensive forts of Washington

Capt. Linus R. Clark.
US Army Heritage Education Center.

us. In March Capt. Steves[1] resigned and went home. Linas Clark[2] took command of Co. F in March.

Wed Apr 1st 1863. Camp near Fort Ripley. The weather is cold and windy, in the afternoon we worked on the Fort. There is only one Sergt that goes on the work each day, that brings us on once in four days.

Thurs. Apr 2nd Today it has been pleasant. I have been in camp all day.

Fri. Apr 3rd. I went with ten men to get some trees to set around our camp it is pleasant.

Apr 4th 1863. To-day the wind blows hard we are all in camp.

Sunday Apr 5th. This morning the snow is 8 inches deep. It has thawed all day. We did not have inspection or dress parade.

Apr 6th. In the forenoon we cleared the snow from the street, in the afternoon I went to work with the men.

Apr 7th. To-day it has been quite pleasant.

Apr 8th. At six o'clock we had dress parade.

Apr 9th. Charlie and myself went to Washington. We went to the Surgeon General's office, and he examined Charlie's arm. Then we went to the Patent Office then around the City until 3 o'clock P.M. and then started for camp. We arrived at camp at sundown.

---

1 Seth J. Steves from Oriskany had prior service as second lieutenant in the 81st New York. He was discharged March 3, 1863, due to illness.

2 Born in Ireland, Linus R. Clark was a farmer before enlisting. On Oct. 27, 1864, he was seriously wounded at Darbytown Road, VA, necessitating amputation of his leg below knee.

Fri Apr 10th 1863. To-day I am Sergt of the guard, it is very pleasant.

Sat Apr 11th. I came off guard 9 o'clock A.M. In the afternoon we had Battalion drill for one hour. Pleasant.

Sun Apr 12th. At nine o'clock had inspection at noon we had orders to get ready to march. Charlie and I left a box of clothes with L. A. Lodge, Tennallytown D.C. I have in the box one shirt one pair of drawers one scarf one pair rislets [?]. It is pleasant.

Mon Apr 13th. We are all ready for a march but the order has not come yet. It is pleasant.

Tues Apr. 14th. We packed our knapsacks and in the afternoon had Battalion [drill] is pleasant.

Wed. Apr 15th. This morning at 2 o'clock we had revil[l]e. Had orders to pack our knapsacks. At seven we were ready to march. Went to Washington to a boat by the name of Enoch Dean. We pushed off at half past 12 o'clock. Went down the Potomac until dark and then anchored. Rained all day.

In February of 1863, Federal authorities dispatched the Ninth Corps to Fortress Monroe in Hampton, Virginia, on the eastern tip of the York Peninsula across the mouth of the James River from Norfolk. Fearing a possible drive on Richmond, Lee ordered Gen. James Longstreet to take the divisions of Gens. John Bell Hood and George Pickett to protect Richmond, gather forage and supplies to support Lee's army, and if possible drive Union troops out of Suffolk. This caused a further shifting of Union forces to bolster the defense of Suffolk. Special Orders No. 55, issued by the Military Department of Washington on April 11, the same day that Longstreet arrived near Suffolk, designated over two dozen infantry regiments and several batteries of artillery to reinforce the troops holding Suffolk. One of those was the 117th New York.

33d ANNIVERSARY
117th N.Y.V.
New York Mills,
AUGUST 21, 1895.
Honorary.

Campaigning in Virginia

April 15, 1863 — July 30, 1864

The men rose early on April 15 and packed their equipment and belongings. With all in readiness, accompanied by a band playing "The Girl I Left Behind Me," the men marched to the Naval Yard to board the steamers Zeypher *and* Enoch Dean. *After anchoring overnight, they passed Fortress Monroe the following evening but continued on to Norfolk, Virginia, arriving on April 17. After another wait before going ashore, the regiment formed ranks, marched through the city, and boarded a train for Suffolk. As the cars neared their destination, the distant boom of artillery fire brought home to the nervous soldiers the realization that they were entering active campaigning with all its attending dangers.*

Assigned to a brigade under Col. Harrison S. Fairchild of the 89th New York as part of Gen. George W. Getty's Third Division, Ninth Corps, defending Suffolk, the regiment took a position along the east bank of the Nansemond River. As if to emphasize the danger, Col. Pease issued an order reminding the men that it was "of the highest importance that the strictest vigilance be observed on picket and in the rifle Pits. The Safety of the command and the whole line depends upon this and the negligence of one man may Sacrifice the whole. Everything indicating any movement or stir in the enemy's lines should be promptly reported to the commanding officer."[1]

Thurs. Apr 16th. This morning it is cloudy, at 6 we found ourselves at Acquia creek. We started off at noon. We lost sight of Maryland on our left. We can see the Virginia shore. The sun is shining and is very pleasant. At ten o'clock P.M. we passed Fortress Monroe.

Fri Apr 17th. This morning we find ourselves at Norfolk. At seven o'clock A.M. we took the cars for Suffolk, arrived there at ten o'clock A.M. we drawed our shelter tents and marched through Suffolk and 1 mile out of the place and pitched out tents, at dark we had orders to march. We left our tents and knapsacks. We marched about 2 miles and was formed into line of battle at one o'clock P.M. We slept on the grounds. Pleasant and warm.

---

[1] Order No. 14, 117th NY Order Book, RG 94, NARA; Steven A. Cormier, *The Siege of Suffolk: The Forgotten Campaign, April 11-May 4, 1863* (Lynchburg, VA: H.E. Howard, 1989), 121, 312.

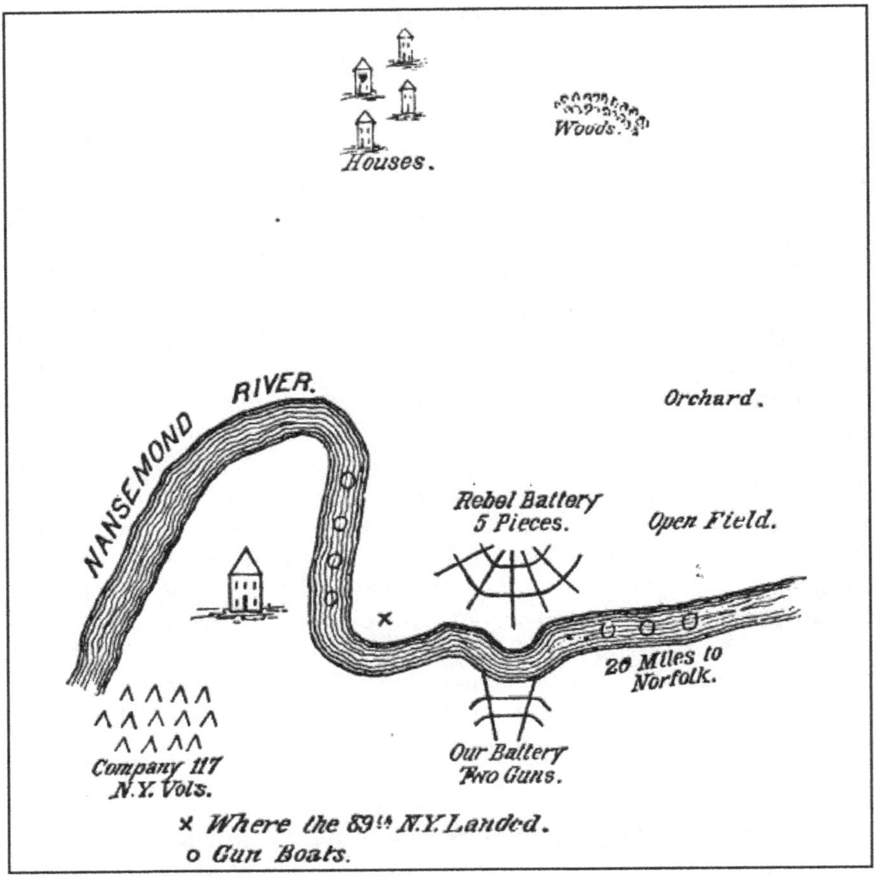

Drawn by John Knox Williams, an officer in the regiment, this shows the position of the 117th New York along the Nansemond River across from a Confederate battery.

Sat Apr 18th. This morning we found ourselves all right. We are in line of battle yet. At 10 A.M. we marched. We went four miles down Nansemond river. There has been one hundred men detached out of our Regt to throw up rifle pits. The rest of the Regt slept on the ground. It is pleasant.

Sun Apr 19th. This morning our tents and knapsacks came to us at nine o'clock A.M. Our gunboats commenced shelling the rebel battery. Some of our men are in the rifle pits. At 11 o'clock A.M. both sides are firing very steady. All is quiet at 1 o'clock P.M. At 4 P.M. we had orders to get ready to march about this time the 89th Regt crossed the river and took a rebel Battery with six pieces of Artillery and 3000 lbs of powder. We marched up the river 3 miles and laid all night.[1]

Mon Apr 20th. At 3 o'clock A.M. we started for the boat we went across the river about day break there was 1 dead rebel in the battery. We burned a few houses and took 1 prisoner. We were shooting at the rebels all day. At 4 o'clock P.M. we had orders to come on this side of the river. We did so and brought every thing with us. I got in the camp about 8 o'clock. At dark it commenced to rain.[2]

Tues Apr 21st. To-day it is cloudy. We have been in camp all day.

Wed. Apr 22nd. Last night the rebels threw a shell into our camp but it did not explode. We have laid in camp all day it has been pleasant.

Thurs. Apr 23rd 1863. To-day we have had a good many showers, we rec[d] our pay. I received Fifty two 82/100 Dollars.

Fri Apr 24th. To-day it is rainy. I sent Forty Dollars home by Platner the sutler.

Sat. Apr 25th. This morning at 4 o'clock A.M. we were called into line and staied until daylight. Then we went to bed again. It is pleasant today.

Sun, Apr 26th. Reville at 3 o'clock and we stacked arms in the street and went to bed, at nine we had inspection. At 11 I was detailed to go out with a working squad. We took our dinner with us. We came in camp at six o'clock P.M. it has been very warm today.

Mon. Apr 27th. I have been in camp all day. Pleasant.

Tues Apr 28th 1863. Reville at 4 o'clock. We had orders to pack knapsacks. At 8 we marched one mile to a piece of pine woods. Here we pitched our tents. It is a nice camp. We have had all kinds of weather to-day.

---

[1] A portion of the 117th New York supported the attack of the 89th New York.

[2] The companies that crossed the river were led by Capt. David Magill of New York Mills.

Wed. Apr 29th. To-day it has been cloudy. At six o'clock P.M. we had a heavy thunder storm.

Thurs Apr 30th. In the forenoon it rained. We were mustered for 2 months pay. The rebels threw some shells into our camp, but did not hurt any body, it was pleasant in the afternoon.

Fri. May 1st 1863. To-day I am on guard. It is pleasant.

Sat. May 2d. I came off guard at 9 o'clock, been in camp all day, it has been pleasant.

Sun. May 3d 1863. At nine o'clock we had inspection. At 2 P.M. we had orders to march. We went down to the river. 3 companies went across.[1] Our company staid on this side of the river all night. It has been pleasant & warm.

May 4th. This morning a[t] 9 o'clock we were relieved and came to camp. Pleasant.

Tues. May 5th. At five o'clock P.M. we went on picket.

Wed May 6th. Last night it rained most of the time. In the morning we came over to come over to the camp. It has been muddy all day.

Thurs, May 7th. It has rained we have been in campy all day.

Fri May 8th. To-day I rec^{d} my receipt of the sutler for the money he expressed for me at Washington.

Sat. 9th. Five companies went and fixed a new camp ground. It is pleasant. We went in our camp in the afternoon.

Sun. May 10th. No inspection or dress parade to day. Has been pleasant.

Mon, May 11th. This morning we had dress parade and orders to stick tents at nine and move to our new camp which is one mile from here. Very warm.

---

[1] When Col. Arthur H. Dutton of the 21st Connecticut appealed for assistance, Major Rufus Daggett personally led Companies E, H and K into some of Dutton's rifle pits to assist in their defense.

Tues May 12th. We are in our new camp in an open field. At six o'clock P.M. we had dress parade.

Wed. May 13. I have been in camp all day. At six P.M. we had dress parade. Very warm.

Thurs May 14th 1863. This morning at seven we had dress parade. We stacked our guns and left them in line. Had company drill one hour. At six P.M. dress parade. Very pleasant.

Fri May 15th. This morning we had orders to pack knapsacks and be ready to march at seven, we were ready, but did not start until 10. We went towards Suffolk until we came to the Rail Road which is 4 miles. Here we took the cars and went to Suffolk and then we went towards Norfolk. The cars came to a halt and we got off. We marched about 1 ½ miles and then pitched our tents. It has been pleasant.

Sat May 16th. This morning I found myself all right. It is pleasant.

Sun May 17th. This morning we had general inspection. Today Owen Jones[1] was taken sick and went to the Hospital. I took care of him.

Mon May 18th 1863. In the morning it was cloudy and sprinkled at nine. It is pleasant. We fixed our street today.

Tues My 19th. Today I made a bedstead for Owen. It is pleasant.

Wed. May 20th. In the forenoon we fixed the streets, in the afternoon I was detailed to help make bedsteads for the hospital.

Thurs May 21st. I have been in camp all day it is pleasant.

Fri May 22. Last night I set up with Owen Jones. It has been pleasant to-day.

Sat May 23d. In the forenoon we went out to work on the Fort, in the afternoon staid in camp. At six P.M. we had dress parade.

Sun May 24th. We had street inspection at nine and dress parade at six P.M. it has been pleasant.

---

[1] Born in Wales, Owen Jones was a farmer from Trenton. He later served much of the war assigned to regimental and brigade headquarters.

Mon. May 25th 1863. This morning at half past 5 we went to work on the Fort. Worked until 9 o'clock A.M. and in the afternoon from 3 until six. Pleasant.

Tues May 26. Today I am Sergt of the guard. It is cloudy.

Wed May 27th. At nine I came off guard, pleasant.

Thurs May 28th. This morning we went to work on the Fort. I am to work with the Carpenters. At seven P.M. we had undress parade.

Fri May 29th. Today we have been to work on the Fort. It is pleasant.

Sat May 30th. In the forenoon worked on the Fort in the afternoon staid in camp.

Sun May 31st. We had inspection at 9 A.M. Dress parade at 6 P.M. Pleasant.

June 1st 1863. I went out with the Co. to work on the Fort. It is pleasant.

June 2d. I went out with the Co. to work on a gun bed, pleasant.

Wed May 3d [actually June 3]. We went to work at 9 o'clock A.M. rained all day.

Thurs June 4th. In the forenoon we worked in the afternoon we rec[d] our pay. I received Thirty four Dollars.

Fri June 5th. To-day I am on picket. Cloudy.

Sat June 6th. This morning I came off picket. We have drawn wedge tents. Pleasant.

Sun June 7th. This morning we had street inspection and at 6 P.M. dress parade.

Mon June 8th. Today I have been to work with the Co. it is pleasant.

Tues June 9th 1863. To-day we worked on the Fort. I sent home by express $25.00. Pleasant.

Wed. June 10th. Worked on the Fort today. Pleasant.

Thurs June 11th. Worked on the Fort, it is cloudy.

Fri June 12th. Worked on the Fort today.

Sat June 13th In the forenoon worked on the Fort. I have charge of the Co. this week while they work on the Fort. At 6 we had dress parade.

Sun June 14th. This morning we had street inspection and dress parade at six P.M.

Mon June 15th. We worked on the Fort today.

Tues June 16th. Today we worked on the Fort. Pleasant.

Wed. June 17th. We worked on the Fort today, after work I went to a PicNic of the 89th N.Y. Regt.

Thurs. June 18th. Today I have been on the works, been pleasant.

Fri June 19th. I am Sergt of the pickets today.

Sat June 20th. At nine I came off guard. It is pleasant.

Sun June 21st. We have got marching orders. We had dress parade at 6 P.M.

While the 117th New York helped defend the Suffolk area, including lengthy assignments as construction labor, Gen. Robert E. Lee began his second invasion of the North that initiated the Gettysburg Campaign. Although not in a position to confront Lee's invasion directly, the Confederate movement north greatly influenced the future of the Oneida County regiment. Anxious to discomfort Lee's invasion, General-in-Chief Henry Halleck, himself a native of Oneida County, ordered all available troops to "threaten Richmond by seizing and destroying their railroad bridges over the South and North Anna Rivers, and do them all the damage possible." This threat, it was hoped, would "occupy a large force of the enemy" that might otherwise be available for Lee's use.[1]

---

1 Letter, Dix to Halleck, Dec. 15, 1863, *Official Records*, Vol. 46, Pt. 1, 18.

The following entries relate to the movement of the regiment toward Richmond and its subsequent withdrawal to Portsmouth once the campaign concluded. It was a brutally hot time, with the sometimes forced marches leading to sunstroke and even some deaths in the regiment.

Mon June 22d. This morning we had reville, at half past 2 we marched, at 5 we arrived at Portsmouth, at seven we took a boat by the name of Hero, past Fortress Monroe at nine P.M. we went up the York river to Yorktown, landed, camped on the ground that Cornwallis surrendered to Washington.

Tue. June 23d 1863. Today we lay still. Pleasant.

Wed June 24th. At 8 o'clock we had Battalion drill one hour, the rest of the time lay in camp. Pleasant.

Thurs June 25th. Last night it rained. It is cloudy and rained in the afternoon.

Fri. June 26th. Reville at 3 o'clock A.M. at 4 we marched to the landing, took a boat by the name of John Brooks, went up York river to West Point then went up the Pamunkey river and landed at the White house. Arrived at 2 o'clock P.M. pitched our tents tore down rebel houses to get boards to sleep on, it is pleasant.

Sat June 27th. Last night it rained. At 2 P.M. we had inspection. At 7 we had dress parade. It is pleasant.

Sun June 28th. Today it has been cloudy. There is seventy thousand troops here now. At six o'clock we had dress parade.

Mon June 29th 1863. At 3 o'clock P.M. we had Battalion drill, it has been cloudy.

Tues June 30th. At ten o'clock were mustered for 2 months pay. Pleasant.

Wed July 1st 1863. Reville at 4 o'clock, at 6 we were ready to march. We crossed the R. R. bridge that runs over the Pamunkey river. We marched until nine A.M. then stopped to eat dinner we rested 1 hour we then marched [to] King William's Court house and camped in an open field. We arrived here at 3 o'clock P.M. To-night I am Sergt of the

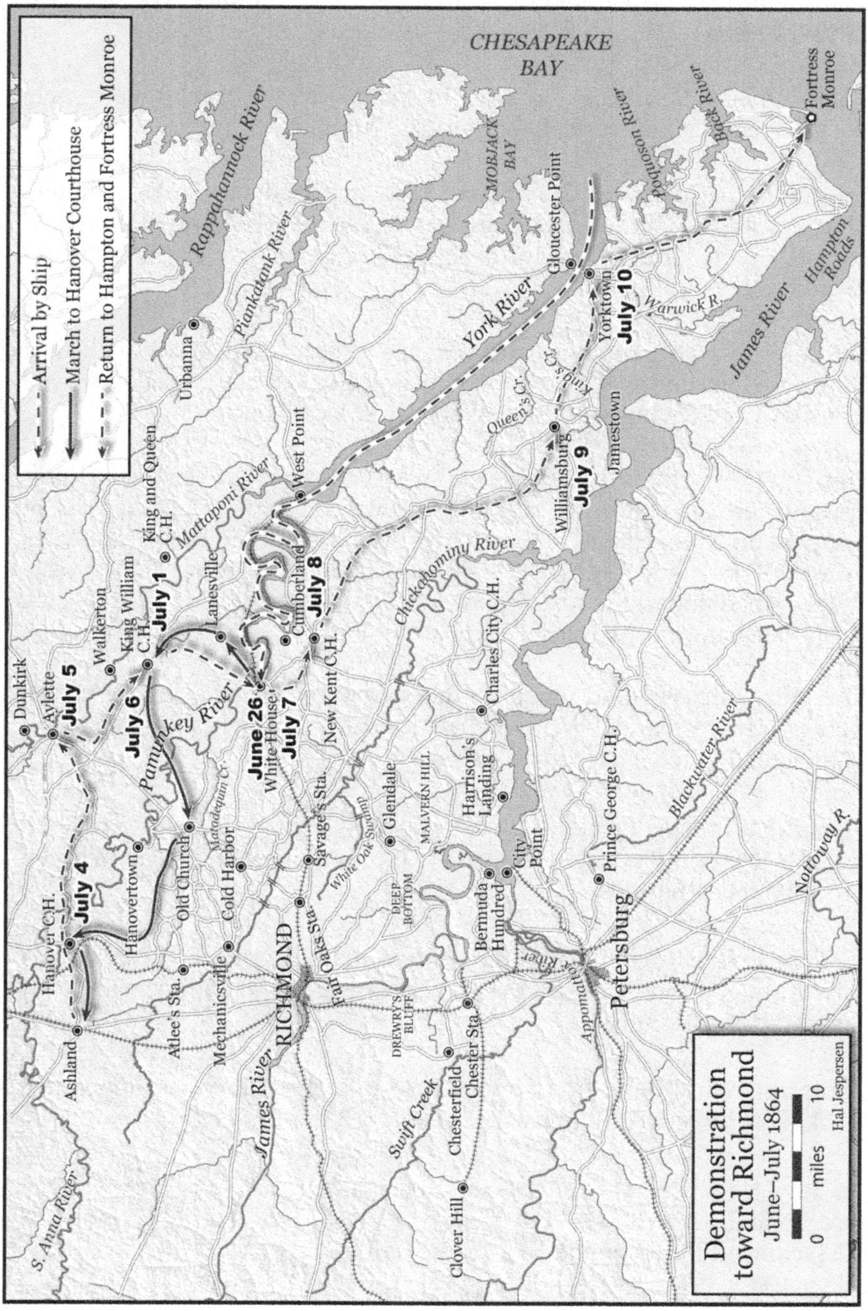
Demonstration toward Richmond
June–July 1864
0 miles 10
Hal Jespersen
Arrival by Ship
March to Hanover Courthouse
Return to Hampton and Fortress Monroe
CHESAPEAKE BAY
MOBJACK BAY
Rappahannock River
Piankatank River
Mattaponi River
Pamunkey River
York River
Poquoson River
Back River
Warwick R.
James River
Hampton Roads
Chickahominy River
Blackwater River
Nottoway R.
Appomattox River
Swift Creek
S. Anna River
Matadequin Cr.
White Oak Swamp
Queen's Cr.
King's Cr.
Fortress Monroe
Gloucester Point
Yorktown
July 10
Jamestown
Williamsburg
July 9
West Point
Urbanna
King and Queen C.H.
Cumberland
July 8
New Kent C.H.
White House
June 26
July 7
Lanesville
King William C.H.
July 1
Walkerton
July 6
Dunkirk
Aylette
July 5
Hanovertown
Old Church
Cold Harbor
Hanover C.H.
July 4
Ashland
Atlee's Sta.
Mechanicsville
RICHMOND
Fair Oaks Sta.
Savage's Sta.
Glendale
MALVERN HILL
DEEP BOTTOM
DREWRY'S BLUFF
Harrison's Landing
Charles City C.H.
Bermuda Hundred
City Point
Prince George C.H.
Petersburg
Chester Sta.
Chesterfield
Clover Hill

guard. Our men drove the rebels from here this morning. We marched 14 miles today.

Thurs July 2d. Reville at 2 o'clock A.M. at 4 we were ready to march. We camped in an open field. Very warm. Marched 11 miles.

Fri July 3d. Reville at 2 o'clock A.M. marched at 4 went through Mechanicsville. We marched until ten o'clock A.M. At night we could not get any water until we got to this place. Been very warm, marched 18 miles.

Sat July 4th. This morning we marched at ten o'clock we arrived at Hanover Courthouse at 6 P.M. our Co. and Co. D went on picket we are 15 miles from Richmond. Marched 8 miles. Very warm.

Sun July 5th. At day break we had orders to report at Hanover Courthouse where the rest of the Regt was, then had orders to retreat and not bring on an engagement. We marched back burned the bridge that runs over the Pamunkey river. It is 1 mile this side of Hanover Courthouse. We marched until noon rested until 5 P.M. then marched until 2 o'clock A.M. then camped. Marched 33 miles.

Mon July 6th. We find ourselves near Elliot's station. At 5 A.M. we were ready to march. We marched to King William's Court house. Marched 10 miles, I has rained to-day.

Tues July 7th. Reville at 4 o'clock. Marched through Leonardsville and back to the White house. Camped on the same ground we did a week ago. Marched 10 miles.

Wed July 8th. Reville at 4 A.M. at five ready to march. Marched by New Kent courthouse, it has rained all day and the muddiest road I ever saw. Marched 10 miles.

Thurs July 9th. Reville at 4 o'clock A.M. marched all day. Camped in a blackberry lot near Williamsburg. Marched 23 miles.

Fri July 10th 1863. Reville at 3 o'clock marched to Williamsburg before breakfast. Williamsburg is quite a place. Arrived at Yorktown 6 P.M. camped for the night. Marched 15 miles.

Sat July 11th. Today we have lain still with orders to march in the morning.

Sun July 12th. Reville at 3 o'clock A.M. marched until noon which brought us to Big bethel. We staid all night. Marched 15 miles.

Mon July 13th. Reville at 4 A.M. marched and arrived at Hampton roads at noon. At six P.M. we had orders to march to the landing. Took a steam boat arrived at Portsmouth at nine P.M. then went to an old camp arrived at midnight. We marched 30 miles.

Tues July 14th. This morning we got up when we got ready. Found our Camp the same as when we left it. Had a few showers to-day.

Wed July 15th. Been in camp all day.

Thurs July 16th. Last night we had the heaviest thunder storm since I have been in this country.

Fri July 17th. I have been in camp all day.

Sat July 18th. It has been very warm.

Sun July 19th. Last night Jonathan Letson died,[1] he belonged to our Co. at nine o'clock we had inspection. Formed into line. We expected Major Gen Foster[2] but he did not come, had dress parade and went to our quarters.

Mon July 20th 1863. At nine o'clock was Letson's funeral, he was taken to Portsmouth. At ten we were ordered to fall into line. Today Gen. Foster came. He is a good looking man. We were on parade one hour then went to our quarters.

Tues July 21st. We have been taking trees out of our street.

---

[1] Jonathan Letson was a Westmoreland farmer. He died of exhaustion from the difficult march at Camp Haskins, VA, on July 19, 1863.

[2] Gen. Robert S. Foster was from Indiana. He commanded a brigade in the Seventh Corps.

Wed July 22d. Today I have been in camp all day.

Thurs July 23d. We have been fixing poles to hang our clothes on it is pleasant.

Fri July 24th. I was detailed to go out with the choppers. They chop 3 hours a day. At 8 in the evening we went on picket to deep creek. We arrived there at 10 P.M. and slept in the Meeting house.

Sat July 25th 1863. I am acting orderly Sergt of the 50 men that came down here last night. Cos. F, I, & E came here.

Sun July 26th. At nine we were relieved and went to our Co. we are quartered in a big barn.

Mon July 27th. At four o'clock we had orders to get ready to march to great bridge which is 8 miles. We arrived at ten o'clock P.M. We slept in an old house. There were no troops here until we came.

Tues Jul 28th. Today we have been fixing our Bunks. Ecker, Lewis, & myself have a room together. It rained last night.

Wed July 29th. This morning we were called at 4 o'clock to pack our knapsacks. At 5 we marched. Arrived at Deep creek at 8 A.M. Camp Haskins at 10. The Regt had gone to Portsmouth at 2 o'clock P.M. we stacked our arms in the street with the rest of the Regt. Took our quarters in an old store. Marched 18 miles.

Thurs July 30th 1863. Today we laid in Portsmouth. It is pleasant.

With the Gettysburg Campaign concluded, the 117th New York's division received orders to report to the Union forces laying siege to Fort Wagner and Charleston, South Carolina. For the upcoming campaign, the regiment served in Col. Samuel M. Alford's Second Brigade, of Brig. Gen. Israel Vogdes' division, in the Tenth Corps led by Maj. Gen. Quincy A. Gillmore.

The Siege of Charleston

July 31, 1863 — April 19, 1864

Located on Morris Island along the southern entrance to Charleston harbor, South Carolina, Fort Wagner was a substantial sand and earth fortification reinforced by logs and sandbags. Situated between the Atlantic Ocean and a swamp, it was a formidable defensive work approachable only across an open sandy flatland. On July 11, and again on July 18, 1863, Union troops from the Tenth Corps under Gen. Quincy Adams Gillmore unsuccessfully attempted to carry the fort, the latter attack made famous in the movie Glory *about the African-American 54th Massachusetts Infantry. Despite the reverses, Federal authorities determined to reduce the fortification. For this purpose, the government reassigned the 117th New York and other troops to bolster Gillmore's command. The reinforcements arrive within sight of Charleston in the early afternoon of August 2.*

Fri July 31st. At ten A.M. we had orders to get ready to take the boat, at one we pushed off, went to Fortress Monroe. Here we got a pilot and pushed off again. We had been gone about an hour when we had a very heavy thunder storm which lasted 2 hours, and then it cleared off pleasant. A good many of the boys were seasick.

Sat Aug 1st 1863. Last night we run all night and today at 4 P.M. we halted at Beauford N.C. and changed steamers. We were on the Georgian,[1] now we are on the S. R. Spaulding. At seven o'clock we started for Charleston. Pleasant.

Sun Aug 2d. This morning it is pleasant and the sea is calm. At 4 P.M. we arrived in Charleston. Here we anchored for the night.

Mon Aug 3d. At nine A.M. went down the coast about 8 miles and landed on Folly Island. Then we marched towards Morris Island. We marched four miles and camped for the night.

Tues Aug 4th. Today we have laid still. At six o'clock P.M. we had dress parade. Pleasant.

Wed. Aug 5th. At 5 P.M. we went on picket on Morris Island. We lay in the trenches all night. Pleasant.

---

[1] The name was actually the *Georgianna.* It proved unseaworthy.

Thurs Aug 6th 1863. Today we have laid in the trenches. The rebs have throwed shells at us, and the sharp shooters are to work quite lively. Pleasant.

Fri Aug 7th. We got into camp from picket at one o'clock A.M. at noon I was detailed to go on picket on Folly Island. The island that we are on. Pleasant.

Sat Aug 8th. At nine o'clock A.M. I came of[f] picket laid in came [he means camp] the rest of the day. Pleasant.

Sun Aug 9th. At nine A.M. we had inspection. Pleasant.

Mon Aug 10th. Laid in camp all day. Pleasant.

Tues Aug 11th. Been in camp all day heavy firing on Morris Island. Pleasant.

Wed Aug 12th. At 10 o'clock we had Gen. inspection. Pleasant.

Thurs Aug 13th 1863. This morning it sprinkled a few drops but no rain. At 3 P.M. our Regt went on picket on Morris we were on the outward post about 200 yards from Battery Wagner.

Fri Aug 14th. We are on picket yet. The rebs throw shells at us from Sumpter[,] Johnson & Moultrie,[1] but they have not hit any of us yet. We were relieved at nine o'clock P.M. we got into camp at 11 P.M.

Sat Aug 15th. Laid in camp all day.

Sun Aug 16th. To-day I went to the Doctors and got exempt. I have got 2 biles [he means boils] one on my back & 1 on my <u>butt</u>. Pleasant.

Mon Aug 17th. This morning at 3 o'clock our Batteries opened on Fort Sumpter. I am exempt today. Pleasant.

Tues Aug 18th 1863. It looks like rain.

---

1 Forts Sumter, Johnson and Moultrie were Confederate defensive fortifications protecting the seaward approaches to Charleston. See the map on the next page.

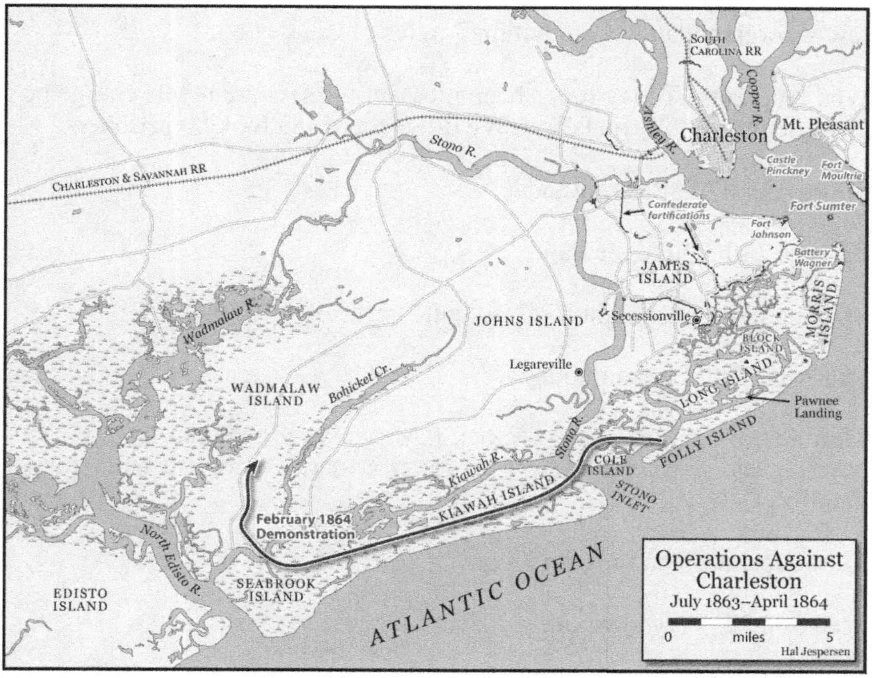

Wed Aug 19th. My biles are well I feel first rate. Been in camp all day. Pleasant.

Thurs Aug 20th. Lay in camp today. Heavy firing on Morris Island. Pleasant.

Fri Aug 21st. At 1 o'clock P.M. we were mustered to see how many men were fit for duty. At 3 o'clock we had inspection.

Sat Aug 22d. In the afternoon we had Battalion drill for 1 hour. Cloudy.

Sun Aug 23d. At ten o'clock I was detailed to go with 10 men to load wood at six we came in camp.

Mon. Aug 24th. In the forenoon we drilled 1 hour, in the afternoon our Regt was detailed to go on picket on Block Island. We went to the head of Folly Island then took a boat went up the Channel landed on Block Island at midnight. It rained all night.

Tues Aug 25th. Today we have been fixing our tents. We have had a few showers today, we are within 3 miles of Charleston.

Wed Aug 26th. Today it has been pleasant we saw the rebels charge on our rifle pits on Morris Island. We drove them and took 80 prisoners.

Thurs Aug 27th. Pleasant today.

Fri Aug 28th. Quite pleasant.

Sat Aug 29th. Last night it rained quite hard.

Sun Aug 30th. Pleasant today.

Mon Aug 31st 1863. At 2 o'clock P.M. we were relieved from picket duty on Block Island. Arrived at camp at dark. Were mustered for 2 months pay.

Tues Sep 1st. Been in camp all day.

Wed Sep 2d. Very pleasant.

Thurs Sep 3d. Today we cleared our street.

Fri. Sep 4th. At nine o'clock A.M. I went on picket. I was on the relieve.

Sat Sep 5th. I was relieved from picket at nine A.M. heavy firing on Morris Is.

Sun Sep 6th. Been in camp all day. Pleasant.

Mon Sep 7th. This morning our men took Battery Wagner & Gregg. Pleasant.

Tues Sep 8th 1863. I have been in camp all day. Pleasant.

Wed Sept 9th. Laid in camp all day.

Thurs Sep 10. Lay in camp today.

Fri Sep 11th. Today we signed the pay rolls for four months pay.

Sat Sep 12th. Today we were paid I rec^{d} \$60. John Ryan[1] was killed on Morris Island.

Sun Sep 13th. At none o'clock I went on picket.

Mon Sep 14th. Came off picket 9 o'clock. Sent home by Adams[2] express Fifty Dollars had it insured.

Tues Sep 15th. Lay in camp all day.

Wed Sep 16th. At 9 o'clock A.M. I went on picket I was on relieve. Pleasant.

Thurs Sep 17th. 1863. At nine o'clock I came off picket. Last night it rained hard.

Fri Sep 18th. It has been cloudy today at 8 o'clock P.M. our Regt were called out to go on reserve. We went about half a mile and lay there until morning then came to camp.

Sat Sep 19th. Lay in camp all day.

Sun Sep 20th. Today it is pleasant at 8 o'clock P.M. we went on reserve about 1 mile from camp.

Mon Sep 21st. At day break we came to camp at noon 100 men were detailed to go on Long Island. We got there about 3 o'clock P.M. I went on picket.

Tues Sep 22d. On picket all day.

Wed Sep 23. At 5 P.M. we were relieved and pitched our tents, we are opposite Seceshville.[3]

Long Island, S.C. Sept 24, 1863. At 7 o'clock A.M. went on picket. Pleasant.

---

1 John Ryan, a farmer from Deerfield, enlisted at age 19 and was killed near Fort Wagner.

2 A reference to the Adams Express Company. Soldiers used it to sent money home to their families.

3 Secessionville. See map on page 23.

Friday Sep[t] 25. At 7 o'clock I was relieved. Pleasant.

Sat Sep 26. Went on picket ab[out] 7.

Sun Sep 27. Relieved from picket.

Mon 28. On picket.

Tues 19. Came off.

Wed 30. On picket.

Thurs Oct 1st. Came off picket.

Fri Oct 2d. On picket.

Sat 3d. Relieved.

Sun Oct 4. On picket.

Monday 5. Relieved from picket.

Tues Oct 6, 1863. On picket, pleasant.

Wed Oct 7. Last night it rained two hours, came off picket 7 o'clock A.M.

Thurs Oct 8. On picket, pleasant.

Friday Oct 9. Came off and went to work building a bridge, worked 4 hours.

Sat Oct 10. On picket at 6 o'clock P.M. being sick with the diahreah, went to the rear and lay with the reserve.

Sun Oct 11th. Last night two Boys from Co. F, one from Co. E, one from Co. H, was caught asleep on the same post that I left yesterday.

Mon Oct 12th. I am exempt from duty today.

Tues Oct 13. Exempt from duty today, fine weather.

Wed Oct 14, 1863. Today I went [with] the working party, to work on the bridge, worked 4 hours.

Thurs Oct 15. Laid on reserve today pleasant.

Fri Oct 16. Went on picket at seven A.M.

Sat Oct 17. At seven o'clock AM came off picket, and ten o'clock A.M. we were relieved by the 13th Indiana Regt. We then came to our old camp on Folly Island. We arrived there at 4 o'clock P.M. and were glad to get back to our Regt.

Sun Oct 18. At 4 o'clock P.M. we had dress parade.

Mon Oct 19. This morning we had a short drill. At ten o'clock A.M. I was subpoenaed to appear at a General court martial, at Gen. Vogdes Head Quarters, at two o'clock PM we were dismissed, and to report there at Ten o'clock the next day.

Tues Oct 20, 1863. At ten o'clock I reported at the court martial, at 2 o'clock was dismissed. The court martial was about the Boys that got asleep on picket at Long Island.

Wed Oct 21. Cos A, B and F are held in reserve; we have orders to get up at 4 o'clock A.M. with equipments on and remain so until daylight. This is a standing order.

Thurs Oct 22. Called up at 4 o'clock A.M. at daylight went to our quarters. At 6 o'clock P.M. was called into line and marched about one mile, and stacked arms. Lay down for the night.

Fri Aug 23d. At sunrise came to Camp: in the fore noon pleasant.

Saturday Oct 24, 1863. Called up at 4 o'clock A.M. and stacked arms until daylight. In the fore noon we drilled one hour, rainy in the afternoon.

Sun Oct 25. Called at 4 A.M. and stacked arms until sunrise, it is quite cold to-day. At six o'clock P.M. went [to] reserve.

Mon Oct 26. At sunrise came to camp, drilled from 11 to 12 o'clock. Pleasant.

Tues Oct 27. At 4 o'clock called up, stacked arms till sunrise, at six PM went in reserve.

Wed Oct 28. At daybreak we came to camp, peasant.

Thurs Oct 29, Called up at 4, stacked arms till daylight, at 6 P.M went on reserve.

Fri Oct 30. At daybreak came to camp. Sergt Lewis Sergt Niles[1] and myself tents together today. Today we bought a stove for five Dollars.

Sat Oct 31st 1863. Called up at 4 o'clock AM, and stacked arms until daylight, at one o'clock P.M. we were mustered for two months pay.

Sun Nov 1st 1863. At four o'clock P.M we had dress parade.

Mon Nov 2d. At four o'clock A.M. we were called up and stacked arms until daylight. At 6 o'clock we went on reserve.

Tues Nov 3d. At sunrise we came to Camp. We drilled one hour in the forenoon, pleasant.

Wed Nov 4th. Called up at 4 o'clock A.M. and stacked arms, drilled one hour in the afternoon. At night went on reserve.

Thurs Nov 5th. At sunrise came to camp drilled one hour in the fore-noon.

Fri Nov 6th 1863. At 4 o'clock A.M. we were called up; stacked arms until sunrise. Drilled one hour in the fore-noon.

Sat Nov 7th. I have been in camp all day, pleasant.

Sun Nov 8th. In the forenoon we had meeting. It is very pleasant today.

Mon Nov. 9. It is quit cold but pleasant.

Tues Nov 10. Today it is very pleasant.

Wed Nov 11. S[t]aid in camp all day.

---

1 Thomas J. Lewis was a wagon maker from Trenton. He was later captured at Bermuda Hundred, VA, on August 25, 1864, and paroled in February 1865. Edgar F. Niles was a carpenter from Oriskany. He was wounded on Sept. 29, 1864, and promoted 1st Sgt. Jan. 1, 1865.

Thurs. Nov 12th. Today we have been fixing the Co. streets. Pleasant.

Friday Nov 13. On picket duty, is pleasant.

Sat Nov 14th. At nine o'clock A.M. I came off picket; in the afternoon we graded the streets. Pleasant.

Sunday Nov. 15, 1863. Today we move our tents to get them into line. Pleasant.

Mon. Nov 16. We have taken the trees out of out streets.

Tues. Nov 17th. I am on camp guard today.

Wed Nov 18. At nine o'clock A.M. came off guard. Col. Fairchilds of the 89th Regt NY. Vols.[1] Inspected our Regt.

Thurs. Nov 19. I have been in camp all day, pleasant.

Fri Nov. 20. I am exempt from duty on account of the erysipelas.[2]

Sat. Nov 21st. Exempt from duty.

Sun Nov 22. Exempt from duty.

Mon Nov 23. Exempt from duty. It has been very rainy today.

Tues Nov 24, 1863. Exempt from duty.

Wed Nov 25. Exempt from duty.

Thurs Nov 26th. To-day is Thanksgiving. At 2 o'clock we had meeting. Pleasant.

Fri Nov 27. In camp today.

Sat Nov 28. Today we signed the pay rolls, it is raining.

Sun Nov 29. Today we got our pay for 2 months. I rec^d Thirty four Dollars, been raining.

---

[1] Col. Harrison S. Fairchild. The 89th was from Binghamton, Mount Morris, Rochester, Norwich, Oxford, Whitney's Point, Windsor, Delhi and Corbettsville.

[2] A bacterial skin infection.

Monday Nov 30th. Very pleasant today.

Tues Dec 1st 1863. I am exempt from duty today.

Wed Dec 2d. Exempt today.

Thurs Dec 3d. This is my birthday. I am exempt. Our Rgt had division drill in the afternoon.

Friday Dec 4th 1863. Exempt from duty.

Sat Dec 5. Exempt today.

Sunday Dec 6th. Exempt. Gen. Vogdes inspected our Brigade. Pleasant.

Mon Dec 7. It is very windy today.

Tues Dec 8th. Very pleasant today.

Wed Dec 9. Our Regt has rec^{d} some apples sent from Oneida Co.

Thurs Dec 10. Exempt today.

Fri Dec 11. Exempt, it has been raining.

Sat Dec 12. Exempt. It has rained very steady today.

Sun Dec 13th. Exempt. Pleasant.

Monday Dec 14th 1863. We drilled one hour this afternoon. Pleasant.

Tues Dec 15. In the afternoon had Battalion drill, pleasant.

Wed Dec 16. Drilled one hour today.

Thurs Dec 17. On camp guard.

Fri Dec 18. At nine o'clock came off guard.

Sat Dec 19. In camp all day. Pleasant.

Sun Dec 20. At 11 o'clock A.M. we had inspection, at 4 P.M. dress parade.

John T. Thomas.
Donald Wisnoski.

Mon Dec 21st. I am on guard at Paume Landing.[1] Pleasant.

Tues Dec 22. At nine o'clock I was relieved.

Wed Dec 23d 1863. In the forenoon we had Battalion drill. Is quiet, cool today.

Thurs Dec 24. At ten o'clock had battalion drill, drilled one hour.

Fri Dec 25. We had an oyster supper and Lieut Thomas[2] treated our Co. with whiskey & segars. Pleasant.

Sat Dec 26. On guard today, there is a cold north wind.

Sun Dec 27. At nine o'clock A.M. came off guard.

Mon Dec 28. It has rained very steady all day.

Tues Dec 29. In camp all day.

Wed Dec 30. Lieut Hunt[3] which was our 1st Lieut has been promoted to Capt in Co. D, 117th Regt and Lieut Thomas has been promoted to 1st Lieut in Co. C.

Thursday Dec 31st 1863. It rained vey hard last night. Lieut Lay[4] of

---

[1] Pawnee Landing. See map on page 23.

[2] John T. Thomas was a Trenton stonecutter. He was later killed in the assault on Fort Fisher, NC, January 15, 1865.

[3] William J. Hunt from Trenton was later the hero of the Union breakthrough at Petersburg Heights, but was wounded during the siege of Petersburg on June 28, 1864, and died of his wounds on July 31, 1864. Hunt Post 510, Grand Army of the Republic, in Holland Patent was named in his honor.

[4] Frank H. Lay of Utica was later captured on May 16, 1864, at Drewry's Bluff, VA, held in Richmond, VA, and Macon, GA, and paroled at Charleston, SC, Sept. 3, 1864. He mustered out with the regiment.

Capt. William J. Hunt.
Donald Wisnoski.

Frank H. Lay.
Oneida County History Center.

Co. I has command of our Co. So ends the year of 1863.

Friday Jan 1st 1864. It is very pleasant, we had for dinner a good oyster soup.

Sat Jan 2d. I am on picket, it is pleasant.

Sun, Jan 3d. At nine o'clock A.M. came off picket. Pleasant.

Mon Jan 4. Very pleasant.

Tues. Jan 5. We had Battalion drill, drilled 2 hours. It has rained some today.

Wed Jan 6. I am on camp guard, rained all day.

Thurs Jan 7. At nine o'clock AM. Came off guard, it has rained all day.

Friday Jan 8th 1864. I have been building a fire place in our tent. Pleasant.

Sat Jan 9. At two o'clock P.M. had inspection.

Sun Jan 10. At 10 o'clock A.M. we had General inspection. Pleasant.

Mon Jan 11. Very pleasant.

Tues Jan 12. In the afternoon we had Battalion drill, is pleasant.

Wed Jan 13. It has rained today.

Thurs Jan 14. On camp guard today.

Fri Jan 15. Came off guard at 9 A.M. pleasant.

Sat Jan 16. Today I am exempt from duty. I have a bile [boil] on my knee.

Sun Jan 17. Exempt from duty.

Mon Jan 18, 1864. Exempt. It has rained all day.

Tues Jun 19. Exempt, very pleasant.

Wed Jan 20. Exempt. The Doctor lanced my knee in 2 places.

Thurs Jan 21. Exempt. My knee is better.

Fri Jan 22. Pleasant as summer.

Sat Jan 23. Exempt. Very pleasant.

Sun Jan 24. Very pleasant.

Monday Jan 25. Pleasant. Exempt.

Tues Jan 26. Exempt: this morning we signed the pay rolls for 2 months pay, in the afternoon I went to the Hospital my knee is quite bad.

Wed Jan 27. I like it very well in the Hospital.

Thursday Jan 28, 1864. Pleasant today.

Fri Jan 29. Our Regt was paid for 2 months. I rec^{d} 34.00 Dollars. Pleasant.

Sat Jan 30. My knee is better. Pleasant.

Sun Jan 31st. To-day I returned to my Company, but not for duty. Knee better.

Monday Feb 1st 1854. Exempt. Very pleasant.

Tues Feb 2d. Pleasant today.

Wed Feb 3d. I sent by Adams express One hundred Dollars to my father. Pike sent $31.00. Griff $10.00 Charley $29.00 and I sent $30.00 which makes $100.00.

Thurs Feb 4th. I am not on duty yet.

Fri Feb 5. Exempt to day.

Sat Feb. 6th 1864. Exempt. Pleasant.

Sun Feb 7. At seven o'clock P.M. our Regt got orders to march. I am not going with them, but my knee is about well. I am exempt.

Mon Feb 8th. We have not heard from the Regt yet. There is but a few men left in Camp.

Tues Feb 9. Today I am on guard.

Wed Feb 10. At 9 o'clock A.M. I was relieved by Sergt Jones[1] of Co. B.

Thurs Feb 11. On guard.

Fri Feb 12. At nine o'clock A.M. was relieved. Our Regt arrived here at three o'clock P.M. all safe and sound.

Sat Feb 13th 1864. On picket, my knee is well. Very pleasant.

Sun Feb 14. At nine o'clock A.M. I was relieved.

Mon Feb 15. Very pleasant to-day.

---

1 John T. Jones, a Utica wagon maker, was later wounded at Drewry's Bluff, VA, and again at Fort Fisher, NC.

Tues Feb 16. Pleasant.

Wed Feb 17. On guard at Pawnee Landing.

Thurs Feb 18. At nine o'clock was relieved. At 4 P.M. had inspection.

Fri Feb 19. Last night it snowed a very little. Just enough to say snow.

Sat Feb 20. It was quite cold this morning but pleasant in the afternoon.

Sunday Feb 21st 1864. At nine o'clock we had street inspection.

Mon Feb 22. Pleasant.

Tues Feb 23. I am on guard at Pawnee Landing.

Wed Feb 24. Our guard were not relieved today.

Thurs Feb 25. At nine o'clock were relieved. Pleasant.

Fri Feb 26. Pleasant.

Sat Feb 27. On picket.

Sun Feb 28. At nine o'clock were relieved.

Mon Feb 29. Very pleasant.

Tues March 1st. On guard at Pawnee Landing.

Wed March 2d. At nine o'clock were relieved.

Thurs March 3d. In camp all day.

Friday March 4th 1864. On guard at Pawnee Landing.

Sat March 5th. At nine o'clock were relieved.

Sun March 6th. In the afternoon we received 35 new recruits.

Mon March 7th. In camp.

Tues March 8. On guard at Pawnee Landing. Our Regt had orders to be ready to march at 2 o'clock P.M. but the order was countermanded.

Wed March 9. At nine o'clock we were relieved.

Thurs March 10th. David Jones[1] died this morning he belonged to Company F.

Fri March 11th. On guard at Pawnee Landing.

Sat March 12th. Relieved at nine o'clock A.M. at 2 P.M. had Regimental Inspection.

Sunday March 13th 1864. In camp.

Mon March 14. On picket.

Tues March 15, At nine o'clock were relieved.

Wed March 16. In camp.

Thurs March 17, To-day I am on camp guard. We received 21 recruits.

Fri March 18. Relieved at nine o'clock A.M.

Sat March 19. In camp.

Sun March 20. On picket.

Mon March 21st. Relieved at nine o'clock A.M.

Tues March 22. In camp.

Wed Mar h 23. In camp.

Thursday March 24th 1864. In camp.

Fri March 25. On picket, been very rainy.

Sat March 26. Rained hard this morning cleared up at nine. I was relieved also.

---

1 Born in Wales, David Jones was a farmer in Trenton. He died of disease at the regimental hospital on Folly Island, SC.

Sun March 27th. In camp.

Mon March 28. On picket.

Tues March 29. Relieved at nine o'clock A.M.

Wed March 30th. On picket.

Thurs March 31st. Relieved at nine o'clock. Windy.

Fri April 1st. Rained in the forenoon pleasant in the P.M.

Sat April 2d. Last night had a heavy thunder storm. On guard at Pawnee Landing.

Sun April 3d. Relieved at nine o'clock.

Monday April 4th 1864. In camp.

Tues April 5th. On picket.

Wed April 6. Relieved at nine o'clock A.M.

Thurs April 7th. In camp.

Fri April 8. On picket.

Sat April 9. Rained very hard last night relieved at nine o'clock A.M.

Sun April 10. In camp.

Mon April 11. In camp.

Tues April 12. On picket.

Wed April 13. At 11 o'clock A.M. were relieved, had orders to be ready to march.

Thurs April 14th 1864. We are under marching orders Cos. A & B came from Long Island last night.

Fri April 15. On guard at Pawnee Landing.

Sat April 16th. Relieved at nine o'clock A.M. signed the pay rolls for 2 months pay.

Sun April 17th. Rec^{d} 2 months pay. Rec^{d} 34.00.

Mon April 18. On picket, at 7 o'clock P.M. was relieved by the 54th Colored Troops.[1]

Tues April 19th. To-day our Regt are all in camp and waiting for Boats to take us. At 5 o'clock P.M. Co^{s} A, B, C, E & I took a Boat and went to Hilton Head.

Wed April 20. At 10 o'clock A.M. had orders to march to Pawnee Landing, got aboard a boat by the name of Sentinel and started off at 4 o'clock P.M. took a schooner in tow loaded with mules.

Gen. Ulysses S. Grant received promotion to head all of the armies of the United States in March 1864. His strategy was to exert maximum pressure on the Confederacy by placing all of the Federal armies in motion at once to prevent the South from shifting its dwindling manpower to meet separate Northern thrusts. The Army of the Potomac under Maj. Gen. George Gordon Meade, with Grant accompanying it, was to move south attempting to bring Lee's Army of Northern Virginia to a stand out in the open where it could be defeated. Maj. Gen. William Tecumseh Sherman would lead an army south from Chattanooga through northwestern Georgia aiming toward Atlanta, destroying railroads and Southern supplies along the way while attempting to neutralize the Confederate army initially led by Gen. Joseph E. Johnston. In Virginia, Grant ordered Maj. Gen. Benjamin Butler to proceed up the James River to attack Richmond.

To reinforcement Butler, Grant established the Army of the James in April 1864 to include Gen. August V. Kautz's cavalry division, Gen. William F. Smith's Twenty-Eighth Corps from North Carolina, and Gillmore's Tenth Corps from coastal South Carolina including the 117th New York assigned to the Second Division under Brig. Gen. John Wesley Turner.

---

1 The 54th Massachusetts Infantry was depicted in the movie *Glory*.

Butler's Virginia Campaign

April 21 — September 28, 1864

As the 117th New York moved to its new assignment, Gen. Grant advised Gen. Butler that he was to advance on Richmond, cut Rebel communications, and prevent reinforcements from being sent from the capital to Lee's army once the Army of the Potomac began its campaign. For the troops arriving from the Carolinas this provided a brief opportunity to rest before the armies took to the roads again. On April 28, Grant informed Butler that he would begin his move south on May 4 and advised Butler to begin his own advance on May 3. Nevertheless, Butler jumped off late, the 117th New York boarding the transport Columbia *for the journey to Bermuda Hundred, located along the James River between Richmond and Petersburg, on May 5.*

Thurs April 21st 1864. We arrived at Hilton Head. This morning at 4 o'clock. At nine o'clock we were transferred to a Boat by the name of Blackstone where the rest of our Regt were. At 2 o'clock P.M. we started from Hilton Head. We had on board our Regt & 2 Companies of the 89th N.Y.V. Weather is fine.

Fri April 22. We sailed all night.

Sat April 23d. This morning we find ourselves all right, at noon we passed Cape Hatteras.

Sun April 24. At six o'clock A.M. we arrived at Fortress Monroe. At nine o'clock A.M. we started for Gloucester Point. We arrived there at 2 P.M. Companys C & F lay on the Boat all night to guard our baggage.

Mon April 25. At six o'clock A.M. came off the boat and struck camp, we are camped at Gloucester Point Va.

Tues April 26th 1864. To-day I sent by express $50 to D. Farley, George Chase sends 20 of it to his wife, and I send the remaining 30. to Darius Farley.

Wed April 27th. Today we had Regimental Inspection. We packed our blankets sent them to Norfolk for storage Henry Lamb went to guard them.

Thurs April 28. To-day we drilled 4 hours.

---

[1] George H. Chase was a painter from Trenton.

Fri April 29. Drilled 4 hours.

Sat April 30. This morning at 7 o'clock we were mustered for 2 months pay, and we had orders to be ready to march at 10 A.M. We went aboard a Boat at 2 P.M. and laid on board all night. Our recruiting party arrived here today with 90 recruits.

Sunday May 1st 1864. At daybreak we started for West Point. Arrived there at noon. I went on picket at 5 P.M.

Mon May 2d. At 5 o'clock P.M. was relieved. At sundown had a very heavy Thunder-storm.

Tues Mary 3d. At ten A.M. we were relieved by Col. Henry.[1]

Wed May 4th. At 6 o'clock P.M. had orders to be ready to march, at one o'clock A.M. we struck tents, and laid around all night.

Thurs May 5. At day break [on a boat] by the name of Columbia went down the York river passed Yorktown at 8 o'clock A.M. passed Fortress Monroe at 11 A.M. then went up the James river.

Fri May 6. At day break we landed at Bermuda Hundred on the James river, at 9 o'clock A.M. we marched 5 miles and pitched our tents.

Sat May 7th 1864. In camp, our men are fighting a few miles from here.

Sun May 8th. At nine o'clock we had Company Inspection, at four P.M. we were called into line, and stacked arms.

Mon May 9th. At 4 o'clock A.M. we were all up and at 5 we were ready to march. We started for the Rail Road that runs from Petersburg to Richmond. We struck the R.R. at Clover Hill Junction, drove the rebels from there, tore up the rail road track and burned all the wood and destroyed all that we could. We then marched towards Petersburg and laid in the woods all night, our men have drove the rebs to day but we have lost a good many men. Our Regt are in the second line of Battle.

---

[1] Col. Guy Henry and his 40th Massachusetts were at that time in the same brigade as the 117th New York. Henry later credited the 117th New York for saving his regiment from being surrounded and captured at the Battle of Drewry's Bluff.

Tues May 10th 1864. The rebels charged on our men 3 times last night but were driven back every time. At 10 o'clock we advanced and took the second line of battle, laid there about an hour and then our men commenced to retreat, and out Regt covered the retreat. We got back to camp at eight o'clock P.M.

Wed May 11th. In camp all day.[1]

Thurs May 12th. Reville at two O'clock at daybreak we marched. We went towards Clover Hill Junction for about three miles then in the direction of the James river. We drove the rebs and a few men in the 3rd N.Y. got wounded and one killed. Company F was on reserve it has rained all day.

Fri May 13th. At nine O'clock A.M. we marched in the direction of Fort Darling and we was in the first line of Battle.[2]

Sat May 14th 1864. Last night the rebs fell back and this morning we follow them up, at seven Oclock P.M. we was relieved from the first line. Our regiment loss to-day is two killed and seventeen wounded. Fred Clark[3] was wounded.

Sun May 15th. All is quiet except the Artillery they kept up a heavy firing all day.

Mon May 16th. At three Oclock A.M. we was called up and stood in line of Battle with orders to charge the rebels works as soon as light enough to see them, but before daybreak the rebs drove our pickets in and charged our lines. They charge on the 142 Regt N.Y.S.V. and was repulsed. The rebs charge three times and by this time there was a general engagement all along our line. We fought until noon and then we

---

[1] With this entry the handwriting in Farley's diary changes radically in its slope and formation, suggesting the rest of the journal was written by someone else. One possibility is that the existing journal was copied from originals and that at this point a different person continues the transcribing. The journal is obviously original and from the Civil War era, but may have been transcribed into a single volume shortly after the war.

[2] Fort Darling was a major fortification along the James River. The entries for May 12-16 refer to the Battle of Drewry's Bluff, Butler's failed attempt to break through the Rebel fortifications guarding the roads into Richmond from the south. The 117th suffered 17 killed, 34 severely wounded, 30 slightly wounded, and eight missing, its largest loss by far to that time.

[3] Charles F. Clark a farmer from Floyd.

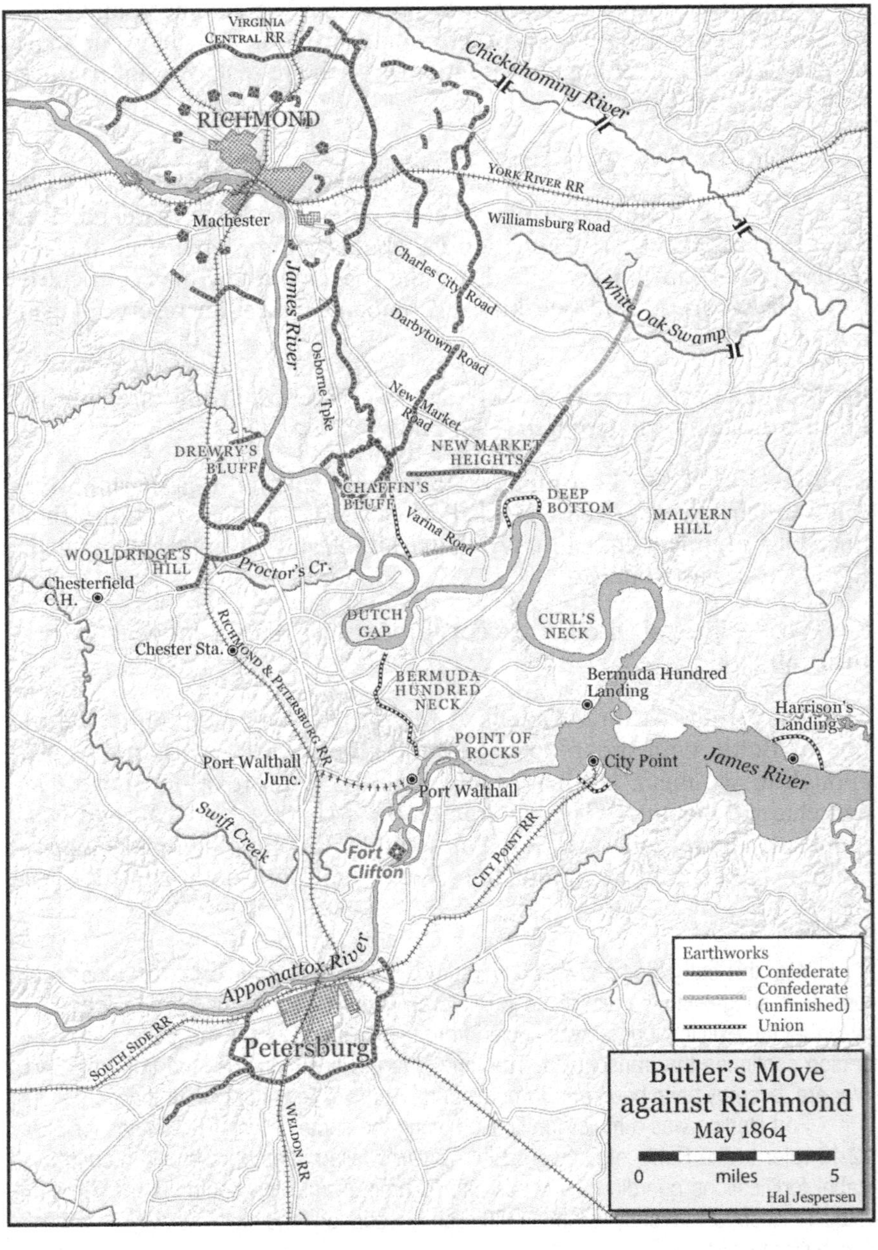
Virginia Central RR
Chickahominy River
RICHMOND
York River RR
Machester
Williamsburg Road
James River
Charles City Road
White Oak Swamp
Darbytown Road
Osborne Tpke
New Market Road
NEW MARKET HEIGHTS
DREWRY'S BLUFF
CHAFFIN'S BLUFF
DEEP BOTTOM
MALVERN HILL
Varina Road
WOOLDRIDGE'S HILL
Proctor's Cr.
Chesterfield C.H.
DUTCH GAP
CURL'S NECK
Chester Sta.
Richmond & Petersburg RR
BERMUDA HUNDRED NECK
Bermuda Hundred Landing
Harrison's Landing
POINT OF ROCKS
Port Walthall Junc.
City Point
James River
Port Walthall
Swift Creek
City Point RR
Fort Clifton
Appomattox River
Earthworks
Confederate
Confederate (unfinished)
Union
South Side RR
Petersburg
Weldon RR
Butler's Move against Richmond
May 1864
0 miles 5
Hal Jespersen

retreated in good order bringing the most of our wounded with us. Our Company left two dead on the field, John McConner and William Wilson. Corp Simons was wounded.[1] Our regt lost in killed and wounded 89 men. We retreated inside the Bermuda fortifications.

Tues May 17th 1864. John Meringer and William David [Davis] died of the wounds they received yesterday.[2] Henry Lamb is detailed as teamster.

Wed May 18th. This morning at One Oclock A.M. part of our regt was call up to work on the fortification. We work until eight Oclock A.M. and then went to camp at ten Oclock we was ordered into line and marched about half a mile and lay there until seven Oclock P.M. then came to camp.

Thurs May 19th. At three Oclock A.M. we was all up and marched about one mile laid there until ten Oclock A.M. and then went to our camp.

Fri May 20th. At One Oclock we was call up into line and stacked arms, at daylight we marched about a mile and halted. At six Oclock P.M. our Regt went on fatigue duty.

Sat May 21st. At seven Oclock A.M. we camp at the place that we halted the day before. We are about half a mile from the rifle pits.

Sun May 22nd. Last night at twelve Oclock the rebs charged our works and was repulsed our Artillery men give them Grape and canister. We are on reserve for the working party.

Mon May 23rd. On reserve for the working party.

Tues May 24th. At seven Oclock we was relieved, in the after noon we laid out a new camp.

Wed May 25th. It has rained some to day. At Six Oclock went on picket. At ten Oclock the pickets commenced firing and had quite a lively time. At day break we went to camp.

---

[1] Born in Ireland, John McConner was an Oriskany farmer. William Wilson was a laborer from Whitestown. Fletcher Simons was a spinner from New York Mills who was discharged on Sept. 10, 1864, due to disability caused by his wound.

[2] Born in Germany, John Merringer was a Deerfield farmer. William H. Davies was a Whitestown farmer.

Thurs May 26th. At five Oclock P.M. we had orders to be ready to march. We are awaiting our further orders.

Fri May 27th. Waiting for orders. At four Oclock in the afternoon we got order to march and went toward Bermuda Hundred about two miles and camped for the night.

Sat May 28th. At two Oclock PM we had inspection and at five we marched. We went across the Appomax [Appomattox] river on the pontoons and then marched towards City Point. At twelve Oclock at night we stacked arms and laid down for the night. We have had several shower to day and the roads are rather slipery.

Sun May 29th. At day break we was called into line and marched to City Point which was one mile. At noon we went aboard the boat Detroit and at seven PM we started down the James river. The weather is fine.

Mon May 30th. Our boat run last night until midnight and then cast anchor. At daybreak we started. Pass New Port News at half past eight. Fortress Monroe at ten. Yorktown at twelve & arrived at West Point at seven Oclock P.M. and cast anchor for the night.

Tues May 31st. At daybreak we started up the Pamonkey river and landed at the White House at eleven Oclock A.M. and then pitched our tents.

Wed June 1st. Our corps has gone for the front except our brigade that is left here to guard the forage. At ten Oclock P.M. we was called up and built a rifle pitt. We got done at twelve Oclock at night.

Thurs June 2nd. Laid still all day. The wagon trains are bringing the wounded in.

Fri June 3rd. At two Oclock P.M. we had marching orders. We marched to Cold Harbor which is sixteen miles. We arrived there at eleven Oclock P.M. and laid in the road.

Sat June 4th. At daybreak we relieved the second line of Battle. Lieut Dann[1] was killed and six of our boys wounded. At seven Oclock P.M.

---

1 Isaac H. Dann from Clinton.

Lt. Issac H. Dann.
US Army Heritage Education Center.

we moved to the right to get in the rifle pitts. It has rained most of the day.

Sun June 5th. Laid in the second line of Battle.

Mon June 6th. At two Oclock P.M. the rebs sent in a flag of truce to bury their dead. It was granted for one hour. At eight Oclock P.M. we was relieved and went half a mile to the rear. David Worden came to see the boys.

Tues June 7th. At eight Oclock A.M. we pitched our tents. At six Oclock P.M. packed knapsacks and struck tents. At eight P.M. we relieved the first line of Battle.

Wed June 8th 1864. Lay in the rifle pitts until eight P.M. and then was relieved and went to the rear about half a mile.

Thurs June 9th. I went to see the 77 Regt and the 146 NYS Vol.[1]

Fri June 10th. At eight Oclock P.M. we relieved the second line of Battle.

Sat June 11th. Laid in the second line, it has been very quiet.

Sun June 12th. Laid in the second line of Battle until ten Oclock P.M. and then we left the line and our corps was orders to march to the White House. We marched all night.

Mon June 13th. At six Oclock A.M. we arrived at the White House and we took a boat by the name of Pawnee and went the Pamonkey and

[1] The 77th New York was from Saratoga County and the 146th from Oneida County (the Fifth Oneida).

York rivers. Arrived at Fortress Monroe at ten P.M. and the anchored for the night.

Tues Jun 14th. At daybreak we started up the James river and landed at Bermuda Hundred at three Oclock P.M. We then marched to our old camp ground at Bermuda Hundred.

Wed June 15th. At one Oclock A.M. we was call up and ready to march. At three Oclock A.M. we cross the Appomax [Appomattox] river on the pontoons and then marched towards Petersburg at eleven Oclock A.M. we took the first line of battle at half past four. Company H and D took a battery of seven guns and ninety prisoners. Our regt was close to these two Company. There was an order for a general charge and we carried the Heights and captured four forts sixteen guns and four hundred prisoners. We had one killed and fifteen wounded. At ten P.M. the second Corps came to reinforce us.[1]

Thurs June 16th. At six Oclock P.M. the second Corps made an attack and we supported them. Our batterys then fire. It was the heavyest canonnading that I ever heard. Our men gained one line of works after two hours hard fighting.

Fri June 17th. At nine Oclock A.M. we moved to the right. At six P.M. our brigade was ordered to Bermuda Hundredth.

Sat June 18th. At three Oclock this morning we arrived at Bermuda Hundredth.

Sun June 19th. At three Oclock P.M. we went up to the fortifications of Bermuda Hundredth and thare all night.

Mon June 20th. Laid in the fortifications.

Tues June 21st. At five Oclock P.M. we moved about half a mile to the left and camped for the night.

---

[1] In the Battle of Petersburg Heights, Capt. William Hunt led Companies H and D through a gap in the Confederate fortifications, catching the Rebels from behind and unhinging the whole defensive line. Hunt's men captured four 12-pounder Napoleons of the Albemarle (Virginia) Artillery, a rifled gun from the Halifax (Virginia) Light Artillery, the colors of the 26th Virginia, and 211 prisoners including the unit's commander, Lt. Col. James C. Councill, two majors, four captains, and nine lieutenants.

Wed June 22nd. Laid inside the fortifications.

Thurs June 23rd. At six Oclock P.M. we had orders to march. We crossed the Appomax [Appomattox] and went towards Petersburg. We had one out of our Company struck with a shell.

Fri June 24th. We arrived at Petersburg Heights at two Oclock this morning and relieved the second line of Battle. At even Oclock P.M. our brigade was ordered into line in front of our works to charge a fort. We formed in line and then the order was countermanded and we relieved the first line of battle.

Sat June 25th. We expected a charge from the rebs last night. They drove in our pickets and then they fell back. It has been warm to day.

Sun June 26th. In the trenches and about three quarters of a mile from Petersburg. The weather is very warm.

Mon June 27th 1864. At six Oclock P.M. the Third NY Regiment[1] reinforced us.

Tues June 28th. At three Oclock this morning the 3rd N Y Regt went to the rear. We drewed some dried apples from the Sanitary Commission. At six Oclock P.M. I went on picket. One man killed and one man wounded from our company.

Wed June 29th. At six Oclock A.M. I was relieved and lay in the trenches the rest of the day.

Thurs June 30th. We was mustered for four months pay. Our brigade got ready to support a charge but there was no charge made.

Fri July 1st. At six Oclock P.M. we was relieved from the first line and we relieved the second line. It has been very quite [quiet] today. One man by the name of Morrell[2] was wounded. He belonged to our Company.

Sat July 2nd. In the second line of trenches.

---

[1] The 3rd New York was from Albany County.

[2] Joseph Morrell, a Utica farmer, enlisted on March 26, 1864, and joined the regiment at West Point, VA, in April. He died of his wounds on September 2, 1864.

Sun July 3rd. At seven Oclock P.M. we relieved the first line. We have four mortars at the left of our Regt and the rebs have three or four opposite. They exchange shell very often.

Mon July 4th. It has been very quite [quiet] until three Oclock P.M. and then there was some shelling on both sides. Our Regt lost one killed and five wounded.

Tues July 5th. At seven Oclock P.M. we was relieved and went to the rear.

Wed July 6th. At seven Oclock P.M. we relieved the first line of rifle pitts.

Thurs July 7th. Our Regiment lost one killed and six wounded by a shell. Henry O'Dwyer[1] was wounded from our Company.

Fri July 8th. 1864. At four Oclock P.M. the rebs made a faint [feint] on our line and we captured eighteen prisoners. At seven Oclock P.M. we was relieved.

Sat July 9th. We are on reserve. It is very pleasant.

Sun July 10th. At seven Oclock P.M. we relieved the first line of rifle pitts.

Mon July 11th. In the rifle pitts.

Tues July 12th. Last night we had a fine shower at seven Oclock P.M. We was relieved. Three wounded out of our regt to day.

Wed July 13th. At seven Oclock P.M. we relieved the second line of breastworks.

Thurs July 14th. At seven Oclock P.M. we relieved the first line of works. I went on picket.

---

[1] Henry Dwyer was born in Ireland and enlisted as a farmer from Trenton. He was wounded at Petersburg July 7, 1864, and again at Chaffin's Farm, VA, on September 29, 1864. He was discharged for his wounds on March 7, 1865.

Col. Alvin White.
US Army Heritage Education Center.

Fri July 15th. At six Oclock A.M. I was relieved from picket.

Sat July 16th. At seven Oclock P.M. we was relieved and we took the second line.

Sun July 17th. I have been with Owen Jones to see Charley and Pike. At seven Oclock P.M. we left the second line and went about half a mile to the rear.

Mon July 18th. At one Oclock P.M. we had inspection. At seven Oclock P.M. we relieved the first line.

Tues July 19th. It has rained hard all day. It is muddy in the trenches. I have quit[e] a sore throat.

Wed July 20th. Last night two rebs came into our lines in front of our Regt. This morning our men and the reb exchanges papers. Samuel Millington[1] wounded. At seven Oclock P.M. we was relieved and went to the rear.

Thurs July 21st 1864. We are on reserve. Col White[2] has resigned and is going home. George Hinman was here today. He belongs to the Sanitary Commission. He distributes some writing paper to the boys.

Fri July 22nd. I went to the camp of the 2nd N Y Heavy Artillery and Dean Rockwell went with me to the camp of the 77th N Y and I got back to camp at four P.M. and at seven Oclock P.M. we relieved the first line.

1 Samuel Millington was probably from Vienna. He was promoted to corporal and assigned as a hospital attendant. He was wounded at Petersburg on July 20, 1864.

2 Alvin White was from Utica. He enlisted as lieutenant colonel and gained promotion to colonel on August 26, 1863. Wounded at Drewry's Bluff, he was discharged due to his wound on July 18, 1864.

Sat July 23rd. in the trenches.

Sun July 24th. At seven Oclock we was relieved and went to the rear.

Mon July 25th. Last night it rained very hard. To day we have been fixing a camp. At four Oclock P.M. we had dress parade and our camp is named camp Curtis in Honer [Honor] of Col Curtis[1] of the 142nd NYV.

Tues July 26th. At four Oclock P.M. we had dress parade.

Wed July 27th. To day I took dinner with Owen Jones. Our Regt is in reserve.

Thurs July 28th. At eight Oclock A.M. I went on guard. At three Oclock P.M. our regiment was inspected [by] Col Curtis and the inspector General. At six Oclock P.M. I was relieved from guard and at seven P.M. we relieved the first line of trenches.

Fri July 29th. At seven Oclock P.M. we were relieved from the trenches and formed a line with our brigade and marched to the left of Burnside line.[2]

Sat July 30th. At five Oclock A.M. one of the rebs forts was blown up our Artillery open and our men charged and took the first line. At seven Oclock A.M. the Colored troops charged and failed to carry the second line. Our division lost in killed and wounded 400 men. At noon we was ordered to the rear.[3]

Sun July 31st 1864. At daylight we had orders to march. We marched through Bermuda Hundredth. At noon we crossed the Appomax on pontoons and halted until three Oclock P.M. we then marched towards the fortifications of Bermuda Hundredth and camped near the General Head Qurters. General Birney[4] Commands the tenth Army Corps. It has

---

1 Col. Newton Martin Curtis commanded the 142nd New York raised in St. Lawrence County. He was later promoted to brigadier general and led the brigade that included the 117th New York in the assault on Fort Fisher, NC, in 1865. The regiment was the first to breach the fort defenses and place its colors on the parapets.

2 A reference to the Ninth Corps led by Maj. Gen. Ambrose Burnside.

3 A description of the Battle of the Crater. The 117th New York was in reserve where it took some artillery fire.

4 Gen. David B. Birney died from disease on October 18, 1864 at age 39.

been very warm. Our division has had seventy men sunstroke to day.

Mon Aug 1st. George Worden has been here to day.

Tues Aug 2nd. We have been fixing our camp.

Wed Aug 3rd. In camp.

Thurs Aug 4th. In camp.

Fri Aug 5th. In the after noon we made a shade over our tents.

Sat Aug 6th. In camp.

Sun Aug 7th 1864. At seven Oclock A. M. we had company inspection at nine A.M. I went on guard.

Mon Aug 8th. At eight Oclock A.M. I was relieved from guard.

Tues Aug 9th. I have been to see Pike.

Wed Aug 10th. In camp.

Thurs Aug 11th. Charley Cookinham[1] has been here. At four Oclock P.M. I went on picket.

Fri Aug 12th. At six Oclock P.M. I was relieved.

Sat Aug 13th. We have orders to be ready to march at four P.M. the most of our Regiment went on picket the rest of our division marched to Deep Bottom and our regiment are left here to do the picket duty.

Sun Aug 14th. At one Oclock A.M. the men that was left in camp was called up and we went up to the trenches. At four Oclock P.M. we went on picket it rained in the after noon.

Mon Aug 15th. At six Oclock P.M. was relieved and lay in a camp near the breastworks.

---

[1] Charles O. Cookingham was a farmer from Trenton. At the time, he was serving as a hospital assistant so he was probably visiting from the hospital. This may have been the Charley that Farley references earlier in his diary.

Tues Aug 16th. In camp.

Wed Aug 17th. At five Oclock P.M. I went on picket.

Thurs Aug 18th. One man was killed out of our Regt by lightening his name was William Wright[1] of Company H. at six P.M. was relieved.

Fri Aug 19th. It has rained all day.

Sat Aug 20th. It was cloudy in the fore noon and it rained in the after noon.

Sun Aug 21st 1864. At three Oclock A.M. our division came back from the north side of the James. We then took over our camp at six Oclock P.M. We had orders to be ready to march at a moments notice with one days rations.

Mon Aug 22nd. At one Oclock A.M. we was call up and marched to the Appomattox river to charge the rebel fort Clifton but the order was countermanded and we marched back and lay in the fortification until sunrise and then came to camp.

Tues Aug 23rd. At nine A.M. I went on camp guard.

Wed Aug 24th. At nine A.M. I was relieved from camp guard. At six Oclock P.M. we marched up to the breastworks and lay down for the night.

Thurs Aug 25th 1864. This morning just at the break of day the rebs charged our picket line and took the right of the line prisoners thirty in number. Sergt Lewis Corpl Conner and Henry Pease[2] was from Co. F.

---

[1] William Wright was a laborer from Vienna. He was on picket duty when a bolt of lightening struck. According to George Fairhead, a member of Company D: "The current of electricity passed along his gun into his left foot & off by his shoulder tearing his shoe and stocking slightly. His body was not torn any. His flesh when brought to camp was a bluish purple." No doubt death was instantaneous.

[2] Thomas J. Lewis, a Trenton wagon maker. Michael Connor was born in Ireland and a Trenton farmer. Henry Pease was a Whitestown carpenter. Lewis and Connor were paroled and mustered out with the regiment. Pease died of disease while a prisoner at Salisbury, NC, Nov. 12, 1864.

Part of our regiment was ordered to take the line that our pickets had but we did so and captured thirty five of the rebs in doing it. We had three wounded in our regiment. At eight P.M. we had a heavy shower.

Fri Aug 26th. We are laying at the breast works. We have got marching orders.

Sat Aug 27th. At six Oclock A.M. we marched for Petersburg, we arrived at Petersburg at six P.M. and relieved the first line of works.

Sun Aug 28th. In the trenches.

Mon Aug 29th. Last night James McCarty[1] was killed by a shell he belonged to Co. F.

Tues Aug 30th 1864. Last night our men shelled Petersburg for about two hours. It is very quiet to day. At four P.M. I went to see Pike.

Wed Aug 31st. At eight Oclock A.M. we was mustered for six months pay. At seven P.M. we was relieved and went to the rear half a mile.

Thurs Sept 1st. On reserve.

Fri Sept 2nd. On reserve.

Sat Sept 3rd. At six Oclock P.M. we relieved the first line of trenches.

Sun Sept 4th. In the trenches.

Mon Sept 5th. Last night we had a heavy thunder storm.

Tues Sept 6th. At six Oclock P.M. we was relieved and went on reserve.

Wed Sept 7th 1864. At six Oclock P.M. I was detailed to go with the working party worked until mid night and then went to camp.

Thurs Sept 8th 1864. On reserve.

---

[1] James McCarthy was a laborer from Utica who had mustered into the regiment on January 4, 1864.

Fri Sept 9th. I have been to see Pike and Simon Wait.[1] At seven Oclock P.M. we relieve the first line of trenches.

Sat Sept 10th. Last night our men gave three good cheers for General Sherman.[2] The cheering commenced on the right of the line and passed to the left.

Sun Sept 11th. In the trenches.

Mon Sept 12th. At seven Oclock P.M. was relieved and went on reserve.

Tues Sept 13th. On reserve.

Wed Sept 14th. Detailed to work on a fort worked all day.

Thurs Sept 15th 1864. At seven Oclock P.M. were sent into the trenches.

Fri Sept 16th. In the trenches.

Sat Sept 17th. In the trenches.

Sun Sept 18th. At seven Oclock P.M. we was relieved and went on reserve.

Mon Sept 19th. At eight Oclock A.M. we had company inspection.

Tues Sept 20th. At eight Oclock A.M. we had General inspection. At seven P.M. our line cheered for General Sheridan.[3]

---

[1] Farley mentions "Pike" frequently. No one with that name, or anything close, appears on the regimental roster. He may have been a friend in another regiment. Simon M. Wait was a Trenton farmer who had apparently been wounded, date unknown, or contracted an illness because he is listed as being mustered out while at a hospital in Wilmington, NC.

[2] The occasion was news that Gen. William Tecumseh Sherman's army had captured Atlanta.

[3] This was prompted by news of Gen. Philip Sheridan's victories in the Shenandoah Valley.

Wed Sept 21st. At daybreak we was aroused by the Artillery which gave a salute for Sheridans Victory at seven P.M. We relieved the first line in the trenches.

Thurs Sept 22nd 1864. It has rained all night. I am detailed to work on the fort. Worked all day.

Fri Sept 23rd. In the trenches.

Sat Sept 24th. At six Oclock P.M. we had orders to be ready to march a[t] eleven P.M. we marched to the tenth corps Head quarters which are on the Heights of Petersburg.

Sun Sept 25th. All is quiet.

Mon Sept 26th. We have fixed our camp.

Tues Sept 27th. In the fore noon I drilled the new recruits. In the after noon we moved our camp a few rods.

While assembling their new camp, the soldiers no doubt anticipated remaining there for at least a few days, but it was not to be. To prevent Gen. Robert E. Lee from shifting forces to confront Gen. U. S. Grant's planned isolation of Petersburg from the south, Grant ordered an assault on Forts Gilmer and Harrison near Chaffin's Bluff north of the James River. The object was, once again, to prevent Lee from shifting forces to oppose Grant's primary advance. To accomplish this, the Eighteenth Corps, under the command of Maj. Gen. Edward O. C. Ord crossed the James to move against Chaffin's Bluff and the Tenth Corps, under the command of Maj. Gen. David B. Birney and reinforced by a division of the Eighteenth Corps, crossed the river at Deep Bottom to advance on Richmond via the New Market and Darbytown Roads. Little could Farley imagine that in less than 48 hours his active campaigning would be over.

Remnant of the colors given to the regiment by the ladies of Utica.
Oneida County History Center.

"I Was Wounded"

September 28 — October 9, 1864

Led by Lt. Col. Francis X. Meyer of Utica, the 117th New York marched on the evening of September 28 to Deep Bottom where it crossed to the north shore of the James River on a pontoon bridge, completing the crossing and going into camp about 2:00 am. After only two hours of sleep, the regiment rose at 4:00 am, cooked coffee and breakfast, and was on the march by 7:30 am. It was in the second line of battle as skirmishers from the 3rd U.S. Colored Infantry drove in the Rebel skirmishers. Once these forward defenses were overcome, the brigade commanded by Col. Rufus Daggett containing the 117th New York formed line of battle to advance. Daggett commanded the 117th before being elevated to brigade leadership.

The brigade advanced under artillery fire that killed and wounded several men. Under increasing artillery and musketry, the 117th New York charged through a small wood, across an open field, and in a final surge the men went up and over the enemy breastworks, the defenders surrendering or falling back in haste. After a brief rest, the men once again formed into line to continue the advance toward Fort Gilmer, a very strongly fortified position along the outer defenses of Richmond held by veteran troops from Lee's Army of Northern Virginia.

We take up Farley's diary with the beginning of the campaign.

Wed Sept 28th. We received orders to be ready to march at three P.M. We marched in the direction of Deep Bottom.

Thurs Sept. 29th 1864. Last night at twelve Oclock we crossed the James river at Deep Bottom. As soon as the break of day we drove the rebs and in the second charge that was made I was wounded. I was struck with a shell in my right hand. I then went to the field Hospital and Dr. Carpenter[1] amputated part of my hand and three fingers. I was then taken to Deep Bottom in an ambulance. At twelve Oclock at night we took a boat for Hampton.

Fri Sept 30th. We are on a boat bound for Hampton Hospital.

---

[1] Henry W. Carpenter was a physician from Holland Patent who served as regimental surgeon.

The U.S. General Hospital in Hampton, VA, had been a female college prior to the war. *Library of Congress.*

Sat Oct 1st. Arrived at Hampton Hospital all sound as could be expected. Wound doing well.

Sun Oct 2nd. In the new camp Hospital section 6th wound doing first rate.

Mon Oct 3rd. A cloudy day and quiet [quite] lonesome but I am feeling very well.

Tues Oct 4th 1864. A very fine morning and everything as lovely as can be expected.

Wed Oct 5th. Wound is doing well.

Thurs Oct 6th. At morn six hundred and eighty wounded soldiers took a boat for New York.

Fri Oct 7th. On the way to New York.

Sat Oct 8th. At daylight we arrived at New York and at noon we took a river boat and landed at Willetts Point fifteen miles from New York.

Sun Oct 9th. I am in ward nineteen. It is a good one and I remained at Grants U.S. General Hospital Willetts Point New York Harbor until December 30th when I received my discharge and at noon on the 31st of December 1864 arrived at Prospect Oneida County New York.

Following his final diary entry from Hampton Hospital on October 3, 1864, because of the severity of his injury Farley received a discharge from further service on December 11, 1864. His comrade-in-arms George Fairhead, who compiled a record of the men in the Fourth Oneida, remembered Farley as a "good man" and "warm hearted comrade" who "was a man & a solger & bore his suffering with fortitude." Returning home, Farley spent the next forty-four years as an active and respected member of his community. Earning a living in the dry goods business, he later engaged in farming and was elected to three terms as supervisor of the Town of Trenton. In 1874 he married Jennie Roberts, with whom he had a daughter who lived a full life and a son who tragically died at age two.[1]

As was the case with many veterans, whose wartime experiences bound them to one another, he quickly became involved in veterans' affairs. After joining Wheelock Post No. 97 of the Grand Army of the Republic in Boonville, he took a leading part in organizing and was a charter member of John T. Thomas Post No. 39 in Prospect, being elected post commander for eight years. A frequent delegate to state GAR encampments, he attended the national encampment in Buffalo, New York, 1901. An active participant in the Oneida County Veterans Association, he was its first vice president, and as a member of 117th Regiment Veteran Association he served as commander in 1888 and attended the Fort Fisher veterans' reunion in North Carolina in Janu-

---

[1] George Fairhead, annotated regimental roster of the 117th New York, Oneida County History Center.

ary 1907.[1]

A Republican in politics and a practicing adherent to the Congregational Church, Farley joined the Trenton Lodge of the Independent Order of Odd Fellows, the Improved Order of Red Men No. 221 in Prospect, the Vallonia Historical Society in which he served as secretary, and the Trenton Association for the Detection of Horse Thieves. When he passed away in Prospect on August 7, 1908, following a long illness, an obituary pronounced his death "an occurrence of unusual sadness. He had made a brave fight for life for several months and bore the long sickness with patience and always greeted his friends with a welcoming smile."[2]

---

1 Wager, *Our Country and Its People*, III, 11; unidentified newspaper clipping at https://www.findagrave.com/memorial/20499672/george-r.-farley.

2 A note on the last page of the diary indicates "Pension No 37823, Registered in Book G, Vol 6, Page 58, By G. L. Ellsworth."

Index

Made in the USA
Monee, IL
05 November 2026

67378966R10042